Energy Calculations &
Problem Solving Sourcebook

A Practical Guide for the
Certified Energy Manager Exam

Energy Calculations & Problem Solving Sourcebook

A Practical Guide for the Certified Energy Manager Exam

Scott Dunning & Larry S. Katz

Routledge
Taylor & Francis Group

NEW YORK AND LONDON

First published by Fairmont Press in 2017.

Published 2020 by River Publishers
River Publishers
Alsbjergvej 10, 9260 Gistrup, Denmark
www.riverpublishers.com

Distributed exclusively by Routledge
605 Third Avenue, New York, NY 10017
4 Park Square, Milton Park, Abingdon, Oxon OX14 4RN

First issued in paperback 2023

Routledge is an imprint of the Taylor & Francis Group, an informa business

Library of Congress Cataloging-in-Publication Data

Names: Dunning, Scott, author. | Katz, Larry S., author.
Title: Energy calculations & problem solving sourcebook : a practical guide for the Certified Energy Manager Exam / Scott Dunning & Larry S. Katz.
Other titles: Energy calculations and problem solving sourcebook
Description: Lilburn, GA : The Fairmont Press, Inc., [2017] | Includes index.
Identifiers: LCCN 2017003858 | ISBN 0881737631 (alk. paper) | ISBN 9788770222594 (electronic) | ISBN 9781138048522 (Taylor & Francis distribution : alk. paper)
Subjects: LCSH: Buildings--Energy conservation--Mathematics--Examinations--Study guides. | Buildings--Energy consumption--Estimates--Examinations--Study guides. | Building management--Examinations--Study guides.
Classification: LCC TJ163.5.B84 D86 2017 | DDC 658.2/6--dc23
LC record available at https://lccn.loc.gov/2017003858

Energy calculations & problem solving sourcebook / Scott Dunning and Larry S. Katz

13:9780881737639   (The Fairmont Press, Inc.)
13:9781138048522   (print)
13:9788770222594   (online)
13:9781003151296   (ebook master)

While every effort is made to provide dependable information, the publisher, authors, and editors cannot be held responsible for any errors or omissions.

ISBN 13: 978-87-7022-943-2 (pbk)
ISBN 13: 978-1-138-04852-2 (hbk)
ISBN 13: 978-1-003-15129-6 (ebk)

Table of Contents

Foreword

In 2008, when my 25+ year career in the telecom industry came to an abrupt close, I decided to pursue opportunities in the energy industry. I learned about energy efficiency equipment and cost measures, electricity and gas generation and distribution, energy procurement, demand-side management, lighting, HVAC, refrigeration, solar, and energy management software. I advised small and mid-size clients to implement or reject energy cost measures by presenting tariff-based cost saving analysis that determined simple payback periods.

To expand my opportunities for professional and personal growth, I opted to attain my Certified Energy Manager credential in 2014. For exam preparation, I studied one of the recommended energy management textbooks from the Association of Energy Engineers website, searched for additional information online across all topic areas, and took copious notes throughout the learning process. I believed my prospects to both pass the exam *and* gain a strong understanding of the material would be enhanced by organizing information in an easily accessible manner to reference during the 4-hour test, and later in my energy consulting work.

The subsequent compendium of energy management information became the basis for *Energy Calculations & Problem Solving Sourcebook—A Practical Guide for the Certified Energy Manager Exam*. Scott Dunning reviewed my compilation of material and suggested that we partner to create a guidebook for individuals studying for the CEM exam. Scott has served as one of the primary course instructors for over 10 years and recently chaired the CEM exam committee through the transition to an ANSI-certified exam.

Our goals for the *Sourcebook* are three-fold:

1. Present an overview of the material for the CEM exam with a guided study that compliments detailed reference material.
2. Provide a reference during the <u>AEE Training Program for Energy Managers</u> seminar and the CEM exam.
3. Be a useful reference throughout an energy manager's professional career.

Practice questions and examples are included throughout the document. While the sourcebook provides a structured approach to prepare for the CEM examination, the authors encourage readers to dive more deeply into the subject matter with a passion for continuous learning and a successful, fulfilling energy career.

Larry S. Katz, CEM, CMVP   *Scott C. Dunning, Ph.D., PE, CEM*
ICF   *University of Maine*

Chapter 1

Codes and Standards

INTRODUCTION

Energy managers need to be aware of applicable codes and standards related to their working environment. While the Certified Energy Manager exam does not cover all possible government codes and standards, it has identified those certain items that it deems important for all energy managers to understand. The list of items are:

Codes and Standards Subject Topics
ASHRAE/IESNA Standard 90.1-2012
ISO 50001
IEC and IEEC Codes
ASHRAE Standard 90.2
ASHRAE Standard 62.1-2010
Model Energy Code
ASHRAE Standard 135-2008

Energy managers will typically maintain a reference library with full documentation on each applicable code and standard. While it is not expected that individuals will memorize complete details, a general knowledge of the background behind each item is expected. We have selected key information that highlights the key aspects of standards and codes.

Recall that **standards** are voluntary guidelines and recommendations established by professional and technical organizations. **Codes** are legal requirements established by governmental agencies. Typically, standards are written in language such that they can easily be included as part of government codes.

1

ASHRAE
(American Society of Heating, Refrigeration, and Air Conditioning Engineers)

1. ANSI/ASHRAE/IES Standard 90.1 - 2012
ASHRAE Energy Standard for Buildings (except low-rise residential buildings).

- US standard that provides minimum requirements for energy efficient designs for buildings *except for low-rise residential buildings.*

- Many states apply the ASHRAE 90.1 standard to different buildings being constructed or under renovation. **Most states apply the standard or equivalent standards for all commercial buildings** while others apply the standard or equivalent standards for all government buildings.

- **Illuminating Engineering Society of North America (IESNA)** and **ANSI** are participating societies.

ASHRAE 90.1 covers:
- Buildings
- Building envelope
- Mechanical and lighting systems (majority)
- New buildings being constructed
- Additions/alterations to existing buildings and their systems

ASHRAE 90.1 does not cover:
- Single family homes
- Multifamily of three stories or less homes
- Manufactured or modular homes
- Buildings that do not use electricity or fossil fuels
- Equipment and building systems that are used for industrial, manufacturing, or commercial purposes

ASHRAE 90.1 Energy Elements Covered:
- **Envelope**
  — Requirements defined by type of building
    - Non-residential conditioned space
    - Residential conditioned space

- • Semi-heated space
- — Insulation
- — Fenestration
- — Doors
- — Air Leakage
- — Roof
- — Walls
- — Floor
- **HVAC**
- **Hot Water**
- **Lighting—lighting power density**
- **ASHRAE/IES 90.1 Lighting Power Allowances** using the Building Area Method

ASHRAE 90.1 Industrial Processes Energy Elements Covered (added in 2013):
- **Building Envelope** include skylights, solar reflectance, thermal emittance, air barriers, and solar orientation
- **Economizers** for data centers (2011)
- **Revisions** affect the maximum:
  - — Fan power limits
  - — Pump head calculation
  - — Chilled water pipe sizing
  - — Radiant panel insulation
  - — Single-zone Variable Air Volume (VAV) Equipment
  - — Supply air temperature reset
- **HVAC**
  - — Energy recovery is required for many more HVAC systems.
  - — Several reheat exceptions were eliminated or modified
  - — Restrictions were placed on overhead air heating
  - — Economizer requirements were added for more climate zones and smaller systems
- **Lighting**
  - — Light Power Densities (LPD) dropped slightly on average
  - — Daylighting and associated lighting control requirements were added
  - — Many lighting control requirements were added, including independent functional testing of lighting controls, occu-

pancy and vacancy controls, exterior lighting controls, and whole-building shutoff

— Offices and computer classrooms now require 50 percent of 120V receptacles to be automatically switched

- Requirements were added for service **water booster pumps** and **elevators.**

- Revised, stricter **opaque element and fenestration requirements** at a reasonable level of cost-effectiveness

- Revised **equipment efficiencies** for **heat pumps**, packaged terminal air conditioners (PTACs), single package vertical heat pumps and air conditioners (SPVHP and SPVAC), and evaporative condensers

- New provisions for **commercial refrigeration equipment** and **improved controls for heat rejection and boiler equipment**

- Improved requirements for expanded use of energy recovery, small-motor efficiencies, and fan power control and credits

- Improved equipment **efficiencies for chillers**

- A new alternate compliance path to Section 6, "Heating, Ventilating, and Air-Conditioning," for **computer room systems**, developed with ASHRAE Technical Committee (TC) 9.9.

The **Federal Energy Policy Act of 2005** established a tax deduction for energy-efficient commercial buildings based on meeting Minimum Requirements set by ASHRAE Standard 90.1-2001. The tax deduction is applicable to qualifying systems and buildings placed in service from January 1, 2006, through December 31, 2007.

A tax deduction of $1.80 per square foot is available to owners of new or existing buildings who implement measures in the following categories that reduce the building's total energy and power cost by 50% or more in comparison to a building meeting minimum requirements set by ASHRAE Standard 90.1-2001.

- interior lighting
- building envelope
- heating, cooling, ventilation, or hot water systems

2. ASHRAE 90.2 is a Residential Energy Standard

Defined for low-rise residential buildings (single family to multi-family).

ANSI/ASHRAE Standard 90.2-2007 - Published standard.
Supersedes ANSI/ASHRAE Standard 90.2 - 2004.
Superseded 90A - 1980 & 90B-1975 for all requirements for low-rise residential buildings

This standard provides **minimum energy efficiency requirements** for the design and construction of:
- **New residential dwelling units** and their systems
- The following where **explicitly specified**:
  — New portions of residential dwelling units and their systems
  — New **systems** and **equipment** in **existing dwelling units**.
  — This standard does not include 'transient' housing such as hotels, motels, nursing homes, jails, and barracks, or manufactured housing.
  — The Standard shall not be used to abridge any safety, health or environmental requirements.

ASHRAE 90.2 standard applies to the following:
- Building envelope
- Heating equipment and systems
- Air conditioning equipment and systems
- Domestic water heating equipment and systems
- Provisions for overall building design alternatives and trade-offs

ASHRAE 90.2 standard does not apply to:
- Specific procedures for the operation, maintenance and use of residential buildings
- Portable products such as appliances and heaters; and
- Residential electric service or lighting requirements.

3. ASHRAE Standard 62.1-2010
Ventilation for Acceptable Indoor Air Quality (IAQ)

Specify minimum ventilation rates and other measures intended to provide IAQ that is acceptable to human occupants and that minimizes adverse health effects.

- Intended for regulatory application to **new and existing buildings, and additions**
- Guide the **improvement of IAQ in existing buildings**
  — Requirements defined for:
    - Ventilation
    - Air-cleaning design
    - Commissioning
    - Installation
    - Operations and Maintenance (O&M)
  — Ventilation requirements based on chemical, physical, & biological contaminants
    - Prescribes a ventilation standard of 15 cubic feet of outside air per building occupant.
    - Level may be ensured by controlling the indoor CO_2 content.
    - Ventilation demand in each zone can be determined by remote CO_2 sensors, similar in manner to a thermostat regulating degree of cooling or heating supplied.
  — In addition to ventilation, the standard contains requirements related to certain sources
  — Additional requirements & other standards may apply for certain spaces (labs, healthcare, industrial, etc.)
  — Not intended to be used retroactively

ASHRAE Standard 62.1-2010 does not include:
- Low-rise residential buildings (found in ASHRAE Standard 62.2)
- Specific ventilation rates for smoking spaces
- Consideration or control of thermal comfort

Acceptable IAQ may not be achieved in all buildings meeting these requirements because of:
- Diversity of sources and contaminants
- Air temperature, humidity, noise, lighting, and psychological/ social factors
- Varied susceptibility of the occupants
- Introduction of outdoor contaminants

Air Changes per Hour (ACH)
In an average HOME, the standard number of air changes per hour

(ACH) = 0.35. In other words, it will take a little less than three hours for the air in the home to recycle entirely.

4. ANSI/ASHRAE/ISO Standard 135-2008-BACnet

BACnet is a *data communications protocol* for building automation systems and control networks.

BACnet was developed per ASHRAE guidance, and now supported and maintained by ASHRAE.

BACnet makes it possible to **integrate a facility's various control systems to a single workstation application** for ease of operation.

Highlights:

- **Flexibility**: allows users to expand and upgrade controls using systems and equipment from multiple vendors
- **Scalability**: can be used to control a few or 1,000's of devices
- Can be used over different data transport networks (Wi-Fi, IP, etc.) accessing multiple locations globally

5. ASHRAE/USGBC/IES Standard 189.1: The Green Standard

Standard 189.1 provides a **"total building sustainability package"** for those who strive to design, build and operate green buildings.

- Site location and sustainability
- Energy use
- Recycling
- Water use efficiency
- Energy efficiency
- Indoor environmental quality

Standard 189.1 serves as a compliance option in the 2012 International Green Construction Code™ (IgCC) published by the International Code Council. The IgCC regulates construction of new and remodeled commercial buildings.

INTERNATIONAL

1. IEC Codes

The International Electrotechnical Commission (IEC) is a non-profit, non-governmental international standards organization that prepares and publishes international standards for all electrical, electronic and related technologies—collectively known as "electrotechnology." IEC standards cover a vast range of technologies including:

- Power generation
- Transmission and distribution
- Home appliances
- Office equipment
- Semiconductors
- Fiber optics
- Batteries
- Solar energy
- Nanotechnology
- Marine energy
- Programmable Logic Controller (PLC) – programming via the **IEC 61131-3 standard**

The IEC charter embraces all electrotechnologies including:
- Energy production and distribution
- Electronics
- Magnetics and electromagnetics
- Electroacoustics
- Multimedia
- Telecommunication
- Medical technology
- General disciplines such as terminology and symbols, electromagnetic compatibility (by its Advisory Committee on Electromagnetic Compatibility, ACEC), measurement and performance, dependability, design and development, safety and the environment.

2. IECC—International Energy Conservation Code from the ICC

Published and maintained by the International Code Council (ICC) as the International Energy Conservation Code® (IECC) as of 1998.

The "**Model Energy Code (MEC)**" *was its predecessor.*

The IECC includes energy efficiency criteria for **new residential and commercial buildings** and **additions to existing buildings**.

The IECC covers the design of **energy-efficient building enve-lopes** and **installation of energy efficient mechanical, lighting and power systems**. The requirements emphasize performance that will result in the optimal utilization of fossil fuel and non-depletable resources in large and small communities.

- **Commercial** buildings
- **Low-rise residential** buildings (3 stories or less in height above grade.)
- Establishes minimum regulations for energy efficient buildings using **prescriptive and performance-related provisions**.
- Fully compatible with all of the International Codes (**I-Codes®**) published by the **International Code Council (ICC):**
  — International Building Code®
  — International Existing Building Code®
  — International Fire Code®
  — International Fuel Gas Code®
  — International Green Construction Code
  — International Mechanical Code®
  — ICC Performance Code®
  — International Plumbing Code®
  — International Private Sewage Disposal Code®
  — International Property Maintenance Code®
  — International Residential Code®
  — International Swimming Pool and Spa Code (TM)
  — International Wildland-Urban Interface Code®
  — International Zoning Code®

IECC is a model code adopted by many states and municipal governments in the United States for the **establishment of minimum design and construction requirements for energy efficiency**.
- Building exterior
- Mechanical systems
- Lighting systems
- Internal power systems

The IECC covers the building:
- Ceilings
- Walls

- Floors/foundations
- Mechanical
- Lighting
- Power systems

3. International Performance Measurement and Verification Protocol (IPMVP)

The International Performance Measurement and Verification Protocol (IPMVP) provides an overview of current best practice techniques available for verifying results of:
- **Energy efficiency**
- **Water efficiency**
- **Renewable energy projects**
- **Indoor Air Quality**

May also be used by facility operators to assess and improve facility performance. **Energy conservation measures** covered herein include:
- Fuel saving measures
- Water efficiency measures
- Load shifting
- Energy reductions through installation or retrofit of equipment
- Modification of operating procedures

4. ISO 50001: Energy Management Systems

Requirements for establishing, implementing, maintaining and improving an energy management system, whose purpose is to enable an organization to follow a systematic approach in achieving continual improvement of energy performance, including energy efficiency, energy security, energy use and consumption.

ISO 50001 requires an organization to demonstrate that they have improved their energy performance.

ISO 50001 Structure:
1. General Requirements
2. Management Responsibility
3. Energy Policy
4. Energy Action Plan
5. Implementation and Operation
6. Performance Audits

7.   Management Review

ISO 50001 Method provides a framework of requirements that help organizations:
- Develop a policy for more efficient use of energy
- Fix targets and objectives to meet the policy
- Use data to better understand and make decisions concerning energy use and consumption
- Measure the results
- Review the effectiveness of the policy
- Continually improve energy management

SAMPLE PROBLEMS

1.   Which consensus standard specifies minimum ventilation requirements?
      a.   ASHRAE 100
      b.   ASHRAE 90.1
      c.   ASHRAE 90.2
      d.   ASHRAE 62

2.   Executive Order 12845 (EO 12845) established energy-efficient acquisition standards for computer-related equipment to meet which EPA requirements?
      a.   Energy Star
      b.   Climate Wise
      c.   EnergyGuide
      d.   EPACT

3.   Which article of legislation was intended to break up the large trusts that controlled the United States electric and gas distribution networks?
      a.   PURPA (Public Utilities Regulatory Policy Act)
      b.   PUHCA (Public Utility Holding Company Act)
      c.   EPACT
      d.   NECPA (National Energy Conservation Policy Act)

4.   What allowed federal facilities to sign long-term performance con-

tracts up to 25 years?

    a.   FEMIA 1988 (Federal Energy Management Improvement Act of 1988)

    b.   EO 12759

    c.   FERC Order 889 (Federal Energy Regulatory Commission)

    d.   EPACT

    e.   ASHRAE 90.1

5.   A 40x30x8 commercial office building has 20 occupants. The outside air ventilation system provides 1.5 air changes per hour. What is the outside air ventilation rate per person for this building, and does it meet the ASHRAE ventilation standard?

    a.   10.1 cfm, No

    b.   12.0 cfm, No

    c.   15.3 cfm, Yes

    d.   20.2 cfm, Yes

    e.   24.0 cfm, Yes

ANSWERS TO SAMPLE PROBLEMS

    1.   a

    2.   b

    3.   a

    4.   b

    5.   a

Chapter 2

Energy Accounting and Economics

INTRODUCTION

One of first lessons an energy manager must learn is that effective management achieves a balance between using the least amount of energy and meeting the firm's capital budget. To accomplish this, one must understand the basics of engineering economics to evaluate the value of a project or piece of equipment to an organization. The exam topics are:

Energy Accounting and Economics Topics

Simple Payback Period	Life Cycle Cost Method
Time Value of Money	Interest Formulas and Tables
Present Worth	Project Life
Net Present Value	Annual Cost Method
Present Worth Method	Economic Performance Measures
After Tax Cash Flow Analysis	Depreciation Methods
Internal Rate of Return	Impact of Fuel Escalation Rates
Energy Accounting	Btu Reporting
Point of Use Costs	Efficiency Measures

SIMPLE PAYBACK

Simple payback refers to the amount of time it will take for a project to pay for itself. For an energy management project, that means how long will it take for the cumulative energy savings to equal the cost of the project. Simple payback is at best a first cut tool for evaluating a project. It should be noted that simple payback does not include the time value of money. It ignores interest, salvage value and project life.

However, many companies still use it for initial evaluation of the cost benefit of a project. The formula is:

PB (Payback) = Initial Cost/Annual Savings

Example
By replacing an older compressor with a newer, more efficient model, a company can save $4,000/year. The cost of the new compressor is $10,000. What is the simple payback on this investment.

Payback = $10,000/($4,000/year)

Payback = **2.5 years**

TIME VALUE OF MONEY

Time value of money refers to the fact that the value of a cash flow is dependent on the point in time it occurs and the relevant interest rate. A cash flow in the present has greater value than a cash flow in the future because today's cash flow can be invested such that its value will have increased at some future date.

If we are given an interest rate for reference, we can relate cash flows over time. Financial terms that we use are:

1. Present Value or Worth (P)—This is the present worth of a cash flow on today's date.

2. Future Value or Worth (F)—This is the future worth of a fixed cash flow or a series of payments at some date in the future.

3. Annual Value (A)—This term describes a series of equal cash flows spaced over time.

4. Interest Rate (I)—This is the rate you would be charged to borrow money. For the CEM exam, this is assumed to be an annual interest rate.

5. Marginal Attractive Rate of Return (MARR)—This is the minimal interest rate that a project must return for a firm to be interested in investing in the project.

INTEREST FORMULAS AND TABLES

To evaluate a cash flow or series of cash flows at a different point in time, we must use either a formula or conversion table. A financial calculator or spreadsheet can provide the conversion factors or one can use interest tables that are provided in an economy text (or as provided with the CEM exam). The conversion factors allow you to calculate the cash flow you want to know, given the cash flow at some different time, assuming a certain interest rate, and the time period between the two values. For example, say that I want to find the future value of a present cash flow five years from now given an interest rate I. My conversion factor would be (F/P,I,N) where N would be 5. Table 2-1 provides a guide to quickly convert from one value to another.

Table 2-1. Financial Conversion Factors

Find	Given	Factor	Other Title for Factor
F	P	F/P, I, N	SPCA (Single Payment Compound Amount
P	F	P/F, I, N	SPPW(Single Payment Present Worth)
F	A	F/A, I, N	USCA(Uniform Series Compound Amount)
A	F	A/F, I, N	SFP(Sinking Fund Payment)
A	P	A/P, I, N	CR(Capital Recovery)
P	A	P/A, I, N	USPW(Uniform Series Present Worth)

Example
An energy saving device saves \$25,000 annually (A) over 8 years. What should a company pay (P) if the MARR [Minimum attractive rate of return] is 15%?

Objective: Solve for P
N = 8 years
A = \$25,000
$(P/A,I\%,N) = (P/A,15\%,8) = 4.4873$
(From the 15% interest table, with N = 8 years)

Annual Interest Rate = 15%
Solve for P = ($25,000/year)(4.4873) = **$112,183 or less**

NET PRESENT VALUE

Often, we find that we have to evaluate the net effect of multiple cash flows or must compare investment alternatives. One tool for doing this is to calculate the **net present value (NPV)**. For an energy savings investment, the net present value of the project is the present value $(PV_{savings})$ of the savings minus the present value (PV_{cost}) of the investment costs. This is a great tool for evaluating projects but can't be used to compare projects with different lifespans. In that situation, it is preferable to evaluate the annual value (A) of each project. This is known as the **Annual Cost Method**.

$$NPV = PV_{savings} - PV_{cost}$$

Example

A project costs $38,250, and will result in $30,500 savings per year for 15 years. What is the Net Present Value (**NPV**) if the interest rate is 10%?

Objective: Solve for NPV
Given values: N = 15 years, I = 10%, A = $30,500/yr
$(P/A,I\%,N) = (P/A,10\%,15) = 7.6061$
(From the 10% interest table with N = 15 years)
First, solve for P = ($30,500/year)*(7.6061) = $231,986
NPV = $231,985 − $38,250 = **$193,735**

INTERNAL RATE OF RETURN

An alternative to net present value analysis is to calculate the **internal rate of return (IRR)**. There exists an interest rate where the net present value equals zero and the present value of the investment equals the present value of the annual savings. This is referred to as the true rate of return on the investment or the internal rate of return (IRR). To solve for IRR, the interest rate (I) is the unknown variable and the other values are given.

Example

A project costs $38,250, and will result in $30,500 savings per year for 15 years. What is the Rate of Return (ROR) which is also the Internal Rate of Return (IRR)?

Objective: Solve for the unknown interest rate (I)

Given values: $N = 15$ years, $P_{costs} = \$38,250$, $A_{savings} = \$11,700/yr$ $P_{costs}/A_{savings} = 3.2692$. We need to find an interest table that provides a factor $(P/A,I,15) = 3.2692$. The 30% table provides the closest value.

So, IRR = 30%.

Example

A project will cost $100,000 today (P), and save $23,400 of energy a year (A) for 12 years.(N). If the MARR is 12%, is this a worthwhile project?

Also, how much money will be available in 12 years if all energy savings are banked (F) and earn 10%(I)?

Step 1: Solve for the interest the project earns and compare it to the MARR of 12%. This is the internal rate of return (IRR) for the project.

Given values: $N = 12$ years, $P = \$100,000$, $A = \$23,400/year$

If you divide P by A, you get $100,000/$23,400 = 4.2735. So, we need to search the interest tables to find one that provides a factor $(P/A,I,12) = 4.2735$. The 20% table provides the closest value which is far greater that the MARR of 12% so the project is worthwhile. **(Calculator yields 21.03%)**

Step 2: Solve for the future value of the investment if the annual savings are invested at 10% interest for 12 years.

Given values: $N = 12$ years, $P = \$100,000$, $I = 10\%$, $A = \$23,400$

We need $(F/A,10\%,12$ years$)$ to determine the future value of the savings $(F/A,10\%,12) = 21.3843$.

$F_{savings} = \$23,400 * 21.3843 = \$500,393$

$F_{costs} = \$100,000 * (F/P,10\%,12) = \$100,000 * 3.1384 = \$313,840$

$FV = F_{savings} - F_{cost} = \$500,393 - \$313,840 = \$186,553$

AFTER TAX CASH FLOW

For companies to evaluate cash flows associated with purchase of a piece of equipment, they must consider that each year the machine gets older and loses market value but no cash flow occurs. The Internal Revenue Service allows companies to count that loss in value against company income. This is known as **depreciation**. Note that depreciation is not a true cash flow but it does allow a company to reduce taxes. There are multiple options allowed to accounting for this depreciation. The simplest is straight-line depreciation.

Straight-line Depreciation = **Total Cost** *of Project/life of project* **years**

After Tax Cash Flow (ATCF) with straight-line depreciation:

ATCF = Annual profit – { (Annual profit – Annual depreciation) x tax rate) }

Annual profit = Annual energy savings, e.g.

Example
A $1.4M project, 10 years, straight-line depreciation, annual savings of $235,000, tax bracket of 34%, has an **ATCF** of:

Depreciation = $1,400,000 / 10 years = $140,000

ATCF = 235,000 – ((235,000-140,000) x .34) = $ 202,700

Note that tax credits are not the same as tax deductions. Tax deductions such as depreciation reduce the taxable income that we are taxed upon. Tax credits are a direct reduction in the amount we pay the IRS.

LIFE CYCLE COST METHOD

Life cycle costing is the process of including all the associated costs over the life expectancy of an asset. If you were to evaluate the life cycle cost of a piece of equipment, you would have to include the initial purchase price, the annual energy savings, the annual energy cost and the salvage value at the end of its life. You could then calculate the net present value of purchasing the piece of equipment. If the net present value is positive, the purchase meets your investment criteria.

Example

Your company is considering purchasing an economizer for your boiler. It will cost $10,000 to purchase it. It will reduce energy costs by $4,000/year. It will last for ten years and have zero salvage value at the end of its life. Your company requires a 20% rate of return on its investments. Consider the life cycle cost of the economizer to determine if it will meet your investment criteria.

$P = A*(P/A,i\%,n)$
$A = \$4,000/year, I = 20\%, n = 10$ years
$(P/A,20\%,10) = 4.1925$
$P = \$4,000*(4.1925) = \$16,770$
NPV = $16,770 – $10,000 = $6,770 (which is greater than zero so invest!)

BENEFIT TO COST RATIO

The **benefit to cost ratio (BCR)** is used to evaluate if the benefits of an investment exceed the costs of the investment. You simply divide the present value of the benefits by the present value of the costs. If the ratio is positive, the project is worth investment.

Example

What is the benefit to cost ratio (BCR) for a project that will cost $100,000, and save $23,400 of energy a year with 10% interest rate for 12 years?

PV Benefits $(A = 23,400/year, I = 10\%, n = 12$ years)
$(P/A,10\%,12) = 6.8137$
PV Benefits $= 23,400*6.8137 = \$159,440$
BCR $= \$159,440/\$100,000 = $**1.59**

ENERGY ACCOUNTING

Facilities use energy that comes in different forms and an energy manager needs to be able to determine the value and cost of each energy form in relation to others. The British thermal unit (Btu) is the common unit used to relate the different energy forms. Conversion units with heating values for common fuels are provided in the Appendix.

POINT OF USE COSTS

When considering switching from one fuel source to another, the energy savings are determined by calculating the energy consumed by the load at the point of use for each fuel source.

Example
A natural gas oven for cooking potatoes currently consumes 3.412 MBtu each month. If you convert the heater to electric, the unit will consume 900 kWh each month. Which approach uses the most energy?

Recall that 1 kWh = 3,412 Btu
900 kWh *(3,412 Btu / 1 kWh) = 3.071 MBtu
Thus, electric energy consumed is less than gas energy consumed.

ENERGY PERFORMANCE MEASURES

Two typical measures for comparing a facility's energy performance are the Energy Use Intensity (EUI) and the Energy Cost Index (ECI). EUI is a measure of the total energy used annually in a facility divided by the area of the conditioned floor space. ECI is a measure of the total cost of energy for a facility divided by the area of the conditioned floor space.

ENERGY USE INTENSITY
(EUI) = Btus/ft^2 per year

ENERGY COST INDEX
(ECI) = $/ft^2 per year

EUI may also be referred to as Energy Use Index.

SAMPLE PROBLEMS

1. To finance a $100,000 energy project at 12% for a 10-year term, what is the annual payment?
    a. $ 8798
    b. $ 17,698
    c. $ 10,558
    d. There is no annual payment; the project will pay for itself.

2. If electricity is selling for $0.08 per kWh and is used in a hot water heater with 90% efficiency, what is the equivalent price of natural gas per therm it gas can be used with an efficiency of 75%?
    a. $1.40/Therm
    b. $1.76/Therm
    c. $2.43/Therm
    d. $2.61/Therm

3. A project's installation cost is $10,000 and will generate $3,235 per year in savings over a 10-year term. Assuming annual cash flows, what is the IRR?
    a. 40%
    b. 35%
    c. 30%
    d. 25%
    e. 20%

4. Which is quickest method to evaluate a project?
    a. Simple Payback Period
    b. Internal Rate of Return
    c. Return on Capital Employed
    d. Return on Assets

5. A 40,000-square-foot building uses the following amounts of energy each year. What is the Energy Use Intensity for this building?

Natural gas	15,000 therms/yr
Fuel oil (140,000 Btu/gal)	5000 gal/yr
Electricity	300,000 kWh/yr

    a. 40,300 Btu/ft² yr

    b.   60,450 Btu/ft²yr
    c.   80,600 Btu/ft² yr
    d.   161,200 Btu/ft² yr
    e.   322,000 Btu/ft² yr

SOLUTIONS

1. b
2. c
3. c
4. a
5. c

Chapter 3

Energy Audits and Instrumentation

INTRODUCTION

Energy auditing is a fundamental practice used to assess energy use in a building, process or system and identify opportunities to reduce consumption. In occupied spaces, energy reductions must not negatively impact human comfort, health and safety, and thus acceptable working and living conditions must be maintained or improved while implementing energy cost measures (ECMs).

The energy manager should be familiar with audit processes, instrumentation, energy assessment techniques, and the equipment in a facility to be evaluated for energy load reduction opportunities.

Topic Areas

Role of Audits	Audit Equipment
ASHRAE Type 1 Audit	ASHRAE Type 2 Audit
Energy Management Measures	Load Factors
Combustion Analysis	Combustion Analyzers
Power Factor Correction	Electric Metering Equipment
Very Basic Thermodynamics	Temperature Measurement
Air Velocity Measurement	Pressure Measurement
Light Level Measurement	Humidity Measurement
Infrared Equipment	Energy and Power Measurement
Fuel Choices	HHV and LHV
Energy Use Index	Energy Cost Index

BASIC AUDIT TYPES

- **Walk-through** energy audit—**least expensive**, entails a **cursory evaluation** of possible energy savings

23

- **Mini-audit**—various **tests and measures are performed** to determine whether particular energy conservation strategies will be effective.
- **Maxi-audit**—most **comprehensive and time-consuming** process
  — Assess amounts of energy consumed in various processes
  — Use computer simulations to identify possible sources of savings

COMMISSIONING

Commissioning—process that ensures a building's systems have been designed, installed, and function according to the needs of the building and its occupants.
- Verifies design and operating parameters have been met
- Reduces building operating costs by maximizing the efficiency of its systems
- Problems may be identified and corrected including design flaws, construction defects, and malfunctioning equipment
- Environmental conditions may be improved leading to increased worker productivity
- Address changes in occupancy and function in buildings that render some systems inadequate

Phases of commissioning:
- **Planning** phase—decisions on which building systems will be analyzed, who will perform the analyses, and how performed
- **Investigation** phase—data accumulated on the selected systems, tests are conducted, test data are obtained, and deficiencies are identified
- **Implementation** phase—correct highest priority problems, improvement in performance is verified
- **Hand-off** phase—deliver report that identifies the improvements and provides recommendations for proper operation and maintenance

Documentation:
- Executive summary of the commissioning effort
- Background information on the building and its systems
- Commissioning plan

- Description of the tests performed and their results, including baseline data
- Description of any deficiencies found and their corrective actions
- Cost savings analysis for the corrected deficiencies
- Recommendations for long term maintenance and operation of the building and any appropriate capital improvements

Continuous Commissioning (also referred to as On-Going Commissioning or Real-Time Commissioning)—a program for *monitoring performance and identifying problems* from existing operation and maintenance programs as well as special tests that are performed as needed.

Initial Commissioning—occurs during production of new building or system, defined as a systematic process beginning with program phase and ending with post-acceptance phase.

Retro-Commissioning—first time commissioning performed in an existing building in which a documented commissioning was not previously performed.

Re-Commissioning—to verify, **improve and document performance** of a building system *that has already had initial commissioning or retro-commissioning implemented in past.* Reapplies the original commissioning in order to keep building system performance in accordance with the design or current operating needs.

ASHRAE ENERGY AUDIT CLASSIFICATIONS

ASHRAE Level I Audit—walk-through audit, starting point for building energy optimization.
Main objective: identify and analyze low-cost/no-cost energy measures

- Initial review of utility bills
- Brief site survey of the building
- Quick assessment of energy and cost saving opportunities
- Identify potential capital improvement opportunities for further investigation

- Provide insight of the building's operation relative to other similar buildings

ASHRAE Level II Audit—a comprehensive building survey and energy analysis.

Main objective: provide a detailed assessment of building systems performance, and identify and analyze most potential energy cost measures.

Building owners, property and operational managers, and tenants contribute their knowledge of the facility including potential areas to repair and/or improve. Audit objectives are defined.

Audit includes:
- Fuel use analysis
- Utility bill and tariff rate analysis
- Energy consumption by system
- Diagnostic testing as necessary based on building performance or operational issues, including:
  — Combustion analysis and steady state efficiency testing
  — Lighting level assessment (foot candles and lighting power density)
  — Blower door testing
  — Duct leakage testing
  — Air flow and temperature measurements
  — Water flow and temperature measures
  — Tracer gas analysis
  — Infrared thermography
  — Solar shading analysis
  — Electrical testing
  — Relative humidity testing
- Energy modeling and/or building simulation
- Scope of work to implement recommendations
- Audit Report

ASHRAE Level III Audit—Extensive instrumentation used for a long term study of the building(s) performance.

Main objective: accurately estimate energy cost savings for complex building or plant operations.

- Long-term trend data collection
- Utilize building energy management systems (BMS)
- Identify ways to optimize operational set-points, sensors
- Identify ECMs
- Identify calibration opportunities
- Predictive modeling

High Level Energy Use Calculations—provide a basic measurement of a building's energy use, and for comparison to buildings of a similar class.

ENERGY BENCHMARK CALCULATIONS

Energy Use Intensity (EUI):
Btus/sq ft/yr—energy in Btus per *conditioned* square foot *per year.*

Energy Cost Index (ECI):
$/sq ft/yr—energy cost in $'s per *conditioned* square foot *per year.*

Load Factor—calculated for a **specified period** (month, hours, day(s), etc.). Indicates the degree of variation in the average amount of electricity required to the highest instance of demand over a period.

$$\text{Load Factor} = \frac{\textbf{Average Load}}{\textbf{Peak Load}}$$

A low load factor indicates that high demand equipment is in operation less often relative to the overall load of a facility; example: a motor that requires 150 kW to cycle on for a short time in a building with an average load of 50 kW.

ENERGY AUDIT INSTRUMENTATION

Electric Metering Equipment
Utility Company Electric Meter is a **Watt-hour meter**.
- Older meters—**count revolutions** of a spinning disc that rotates with a speed proportional to the power consumed.

- Newer meters—**electronic**, display the cumulative product of the **voltage, current, and time on a digital readout**.
  — May also record and display the **power factor and peak demand** with time stamps.
  — Data can be sent to utility by radio or via the power line itself.

A **power factor meter** is used to evaluate the relative phase of voltages and currents (see power factor in Chapter 10). These are typically three phase devices.

$$\text{Power Factor} = \frac{\text{Real Power (actual work performed) Watts}}{\substack{\text{Total Power (Real Power plus Reactive Power)} \\ \text{Volt-Amps}}}$$

Power Meters—used for electrical measurements; many types of portable meters are available from basic single-phase Watt measurements to digital meters with detailed waveform and harmonics analysis.

- Voltage
- Current
- Resistance (Ohms)
- Watts—single and multi-phase
- Frequency
- Power factor
- Harmonics
- Logging functions and data storage
- USB ports for computer access and downloads
- Time stamp
- Min/Max/Average with Time Stamp to record signal fluctuations

Light Measurement

Light Meter or **Foot-Candle Meter**—measures **light intensity**, the **amount of** *illumination* the inside surface a one-foot-radius sphere would be receiving if there were a uniform point source of one candela in the exact center of the sphere.

Illumination is the measure of the **effective visible light** emitted from a device, also known as **Luminous Flux** (luminous power), the measure of the perceived power of light.

1 foot-candle = 1 lumen per sq.ft. = approx. 10 **lux**

Temperature Measurement

Electronic Thermometers consist of **thermistor** or **thermocouple** devices and **pyrometers.**

- **Thermistors are resistors** whose value changes with temperature.
- **Thermocouples are considerably more accurate** and must be used with special wiring. There are several types of thermocouples covering different temperature ranges.
- **Pyrometers** respond to surface temperature or received radiation
    — Some pyrometers convert received infrared energy to an equivalent temperature
    — Optical pyrometers receive incandescent light and quantify the wavelengths to derive the temperature; used to measure **extremely high temperatures.**
- **Bimetallic strip—cheaper alternative**, inaccurate, not recommended for energy use analysis
- **Infrared** thermometer—infers temperature from a portion of the thermal radiation emitted by the object being measured; **finds hot spots and phase imbalances**

Air Velocity Measurement

Air flow measurements are typically used to **ensure proper ventilation or to diagnose problems.** The flow of air is characterized by changes in pressure so in some cases the pressure is what is being measured. This characteristic was formalized by Bernoulli in his principle that stated that as the velocity of a fluid (like air) increases, its pressure decreases.

- **Pitot Tube—pressure measuring** device that is adapted for air velocity measurements according to Bernoulli's principle.
- **Anemometer**—true air velocity measuring device. Types:
    — **Rotating Cup**—cups mounted on horizontal arms to turn vertical shaft; rotations over set time produce average wind speed
    — **Vane**—windmill or a propeller parallel to the direction of the wind
    — **Hot-wire Sensor**—very fine wire electrically heated; air flowing past the wire has a cooling effect on the wire,

changing resistance of wire, to be related to air flow speed
— **Thermal flow meter**—monitors variation in hot-wire temperature
— **Laser Doppler**—use a laser beam of light and produce a Doppler shift for measuring wind speed
— **Sonic**—use ultrasonic sound waves to measure wind velocity.

Pressure Measurements

Typically used to ensure proper airflow in HVAC systems.

Manometer—instrument that **measures pressure**, often using a column of liquid. Manometers may read:

- *Absolute pressure* which is the pressure referenced to an absolute vacuum
- *Gauge pressure* which is referenced to atmospheric pressure
- *Differential pressure* where the difference in pressure between two points is displayed

- Types:
  - **Hydrostatic** gauges (such as the mercury column manometer) compare pressure to the hydrostatic force per unit area at the base of a column of fluid.
  - **Piston-type** gauges—counterbalance the pressure of a fluid with a spring
  - **McLeod** gauge—isolates a sample of gas and compresses it in a modified mercury manometer
  - **Aneroid** gauge—uses a metallic pressure-sensing element that flexes elastically under the effect of a pressure difference across the element.
  - **Bourdon Tube**—common gauge used to measure medium to high pressures using a curved tube that moves based on external pressure vs. tube pressure; the motion is proportional to the pressure to be measured

Inclined manometer is built with the column of fluid at an angle to increase the motion and produce more sensitivity to pressure changes.

Barometer is a scientific instrument used to measure *atmospheric pressure*.

Humidity Measurement

The **hygrometer** is the instrument used to measure humidity. Types:

- **Psychrometer**—evaluates wet bulb and dry bulb temperatures which relate directly to relative humidity.
- **Electronic**—use sensors whose electrical characteristics (usually resistance or capacitance) change with humidity

Infrared measuring equipment

Infrared radiation comes from heat energy and has a wavelength longer than that of visible light. Its *intensity is directly related to the temperature of the surface being measured.*

Infrared measuring devices are primarily used to **identify losses of heat.** An image is generated that shows the intensity of infrared energy with **color variations for different temperatures.**

Modern devices are quite lightweight and portable. Aerial surveys of infrared radiation are available. They provide a good overview of heat distribution external to a building and identify losses including those from external features such as pipelines.

Combustion Analyzers

Combustion efficiency varies depending on the conditions under which fuel is used. Combustion efficiency can be affected by the characteristics of the fuel, the temperature of combustion, and the amount of oxygen and CO_2 present.

Orsat analyzer—measures the relative concentration of individual components of *exhaust gas*—CO_2, CO, and O. Best piece of equipment for **analyzing stack gas** to:

- Identify excessive levels of molecular oxygen and boiler inefficiency.
- Identify excessive levels of carbon monoxide (health hazard)
- Uses a tube of potassium hydroxide, a tube of cuprous chloride, and a tube of potassium pyrogallate; flue gas is introduced to each of these tubes, and the amount of carbon dioxide, carbon monoxide, and oxygen, respectively, removed suggests the extent of presence in the flue gas.
- Nitrogen gas is assumed to remain after the introduction of flue gas.

Although very accurate, this process is time consuming and cumbersome as it requires the use of several wet chemicals within the device.

Fyrite Analyzers—individual analyzers for each gas component; cheaper and smaller than the Orsat type. They are still very accurate and they can be used multiple times without requiring a chemical charge.

Smoke detectors are sometimes used to find unburned carbon in exhaust gases. Measure **wet stack gas vs. dry stack gas** values:
- Chemical cell sensors for O_2 and CO measure on a dry stack gas
- Zirconium Oxide sensors measure the O_2 on a wet and hot stack gas basis.

Measure **Stack Temperature Rise (STR)** with a combination of combustion analyzer, flue gas analyzer, and thermometer.

A Particulate Matter Analyzer measures density of airborne particles including:

- CO and O_2
- Flue gas
- Combustion air temperature
- Flue gas loss
- Flue gas humidity

There are two ways to evaluate the **energy capacity of a fuel**:

- **Higher Heating Value (HHV)** is the full energy content including all products of combustion (in **therms or Btus**, for example)
- **Lower Heating Value (LHV)** *omits the energy in the water vapor that is formed during combustion of the fuel.* The energy of the water vapor is about 10% of the total energy content. LHV values are therefore about 10% less than the HHV value for the same fuel.

Fuel Energy Content
Measurement of **energy density**, the total energy per unit of volume or mass; units in Btu/lb, or Btu/cu.ft., for example.

Harmonic Analyzer

Harmonics are measured using a frequency measurement analyzer, such as a **spectrum analyzer**, to determine **Total Harmonic Distortion (THD)**. A notch filter can be used to eliminate a harmonic.

All electronic equipment creates harmonics and distorts the voltage distributed in a facility.

Certain harmonic frequencies create problems unique to that harmonic:

- **Third harmonic** causes overheating in neutral conductors and transformers.
- **Fifth harmonic** can cause motor issues, such as overheating, abnormal noise and vibrations, and motor inefficiency.

Other typical *non-linear loads* added during energy upgrades include:

- Electronic ballasts
- Computers,
- Controls (PLCs, etc.)
- Various building automation system components

Motor Test Equipment

Stroboscope (or **strobe**) instrument—used to make a cyclically moving object appear to be slow-moving or stationary. The principle is used for the study of **rotating, reciprocating, oscillating or vibrating objects**.

Dynometers measures force, torque, or power of a motor (motor shaft power).

Vibration Analysis is used to find:

- Bad bearings
- Bad gears
- Failing machine mounts

SAMPLE PROBLEMS

1. To measure atmospheric pressure, which of the following may be used?
   a. Bourdon Tube
   b. Pitot Tube
   c. Aneroid Gauge

    d. a and c
    e. a, b, and c

2. Which of the following measures humidity:
    a. Anemometer
    b. Bimetallic strip
    c. Psychrometer
    d. Pyrometer
    e. None of the above

3. An ohmmeter can be used to measure:
    a. Electrical resistance
    b. Pressure
    c. Temperature
    d. Flow
    e. Voltage

4. The most important device on a boiler used to analyze its long
    term performance is:
    a. The pressure gauge
    b. The air/fuel ratio gauge
    c. The stack temperature gauge
    d. The drum water level gauge
    e. There is nothing available to measure the long term perfor-
        mance

5. What type of test instrument would be used to monitor an electri-
    cal system for intermittent power quality problems?
    a. Clamp-on ammeter
    b. Fault meter
    c. Oscilloscope
    d. Chart recorder
    e. Psychrometer

SOLUTIONS TO SAMPLE PROBLEMS

    1. e
    2. c
    3. a
    4. c
    5. d

Chapter 4

Electrical Systems

INTRODUCTION

Electricity is a form of energy that offers the most efficient way to transmit energy over a distance. In facilities, it is used to move energy from one location to another with minor losses occurring during transmission in the form of heat. Energy managers need to understand how to perform basic calculations to determine electrical energy savings opportunities related to when it is used, where it is used and how it used.

The exam topics are:

Electrical Systems

Demand and Energy	Load Factors
Real Power	Reactive Power
Power Factor	Three-phase Systems
Power Factor Correction	Peak Demand Reduction
Rate Structure and Analysis	Motors and Motor Drives
Variable Speed Drives	Affinity Laws (Pump and Fan Laws)
Power Quality	Harmonica
Grounding	IEEE PQ Standard 519

TERMINOLOGY—DEMAND AND ENERGY

Voltage—The volt is a measure of electrical potential difference between two points. It is often envisioned as the "pressure" that drives the flow of electricity. It measures the potential energy available to move a fixed amount of charge.

Voltage (V) = Joules of Energy / Coulombs of electrical charge

Current—The ampere is the primary unit of electrical current. It represents approximately 6.3 billion electrons passing one point in a circuit in one second.

Ampere (I) = Coulombs of electrical charge/second

Resistance—The Ohm (Ω) is a measure of electrical resistance to current flow.

Ohm (R) = Voltage (V)/Current (I)

Power—The Watt is the primary unit for power. Power represents the capacity or *rate* at which work can be performed. In alternating current circuits it is known as **real power**.

One Watt of Power (P) = one Joule/one second
(fairly small number)

Power is usually measured in kW or MW. The rate of power consumption at a facility is called **Demand** or Peak Demand.

Energy—The fundamental unit for energy is the Joule. Energy and Power are related through the variable time.

Energy = Power (P) * Time

Typical units of energy measured by utilities are kWh or MWh.

PEAK DEMAND REDUCTION

Peak demand is a measurement of the highest rate of energy consumption that occurred over a period of time. Utilities typically average peak demand over fifteen minute intervals and the billed demand in a monthly electric bill is the highest fifteen minute average. The process of shifting loads outside of the peak demand period to off-peak or shoulder periods to reduce electric charges is called **peak demand reduction**. This will only save money if a facility is billed using time-of-use metering.

APPARENT, REAL AND REACTIVE POWER

Total (apparent power) (VA) is the product of AC voltage and AC current.

S (Volt-Amperes) = V (Volts) * I (Amps)

Apparent power represents the combination of real power and reactive power. Real power represents the capacity to perform useful work. Elements such as capacitors and inductors do not perform useful work so they do not dissipate power. However, they do impede the flow of current and affect both voltage and current. These elements are known as reactive impedances and reactive power is a function of their measured reactance. Reactive power is measured in Volt-Amperes Reactive (VARs).

L =   inductance and is measured in Henries
C =   capacitance and is measured in Farads

X_c = capacitive reactance = Q/I^2 = $1/(2 * \pi *$ electrical frequency*-capacitance)
X_L = inductive reactance = Q/I^2 = $(2*\pi*$electrical frequency*inductance)
Z =   total circuit impedance = $R + X_l$ or X_c

P =   Real power = $I^2 * R = V^2/R$
Q =   Reactive power = $I^2 * X_{(C\ or\ L)}$
S =   Apparent power = $P + Q$

POWER FACTOR

Reactive elements in a circuit affect the relationship between the voltage waveform and the current waveform. The reactive elements cause the voltage and current to **no longer be in phase**. The phase shift in time that occurs is typically called the phase angle associated with the impedance. Assuming that the current waveform is periodic, the total period of the waveform is broken into 360 equal pieces and each piece corresponds to one degree of a total circle of 360 degrees. The shift in time between the voltage and current waveform or phase shift can then be represented as an angular difference. An inductor causes the current waveform to lag the voltage waveform by 90 degrees. A capacitor causes the current waveform to lead the voltage waveform by 90 degrees. The impedance phase angle is measured as the difference between the voltage angle θv and the current angle θi.

The power triangle is used to illustrate the relationship between the apparent power (S), the real power (P) and the reactive power (Q). It is shown in Figure 4-1. The horizontal axis is called the real axis. The vertical axis is called the imaginary or j axis. The three quantities of S, P, and Q can be related to each other using trigonometry.

$S = P + jQ = V * I$
$P = S^* \text{ (cosine of phase angle)} = V^*I^* \text{(cosine of phase angle)}$
$Q = S^* \text{ (sine of phase angle)} = V^*I^* \text{(sine of phase angle)}$

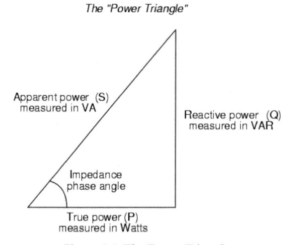

Figure 4-1. The Power Triangle

Power in a direct current (DC) circuit is simply the product of voltage (V) and current (I). No phase shift occurs in a DC circuit. Thus the factor that differs between the calculation for DC power and the calculation for AC power is the cosine of the phase angle shift that occurs in an AC circuit. Thus, power factor is defined as the cosine of the phase angle.

Power Factor = cosine of impedance phase angle = P/S

POWER FACTOR CORRECTION

Commercial and industrial facilities have both resistive and reactive loads. Elements such as transformers and motors contain electrical coils that have significant inductive reactance. If utilities were to just bill for real power consumption (P), it would not capture the true cost of meeting the facility electrical needs. Thus, utilities may bill for apparent power (S) or may bill for real power consumption (P) with a reactive demand charge (Q) for the phase shift caused facility inductive loads. To reduce reactive demand, capacitors can be installed locally. The capacitors act as a local supply for VARs causing the load seen by the utility to be reduced and reducing the phase shift angle. This is known as **power factor correction**.

RATE STRUCTURE AND ANALYSIS

Terminology

    LDC—Local distribution company

    ISO—Independent system operator

    PX—Power Exchange, an open market for the sale and distribution of power.

    EWG—EWGs are exempt wholesale generators, independent power plants that produce electricity for sale in the wholesale market. They are exempt from the restrictions imposed by the Public Utility Holding Company Act that was enacted to control the activities of large power companies.

    Retail and Wholesale Wheeling (de-regulation): customers could buy electricity from a host of competing suppliers who would be given the right to "wheel" their power across the transmission lines of local utilities. Co-ops may be hurt by deregulation because lower costs in attractive regions could mean higher costs in rural areas.

    Primary Service: Voltage service at a transformer prior to step down; customer owns the transformer in this case.

    Secondary Service: The output side of a transformer and the circuit connected with it. Voltage delivered between 0 and 750 volts. Also referred to as service delivery voltage.

    Evergreen Clause: energy contract automatically renews.

Take or Pay: commitment to purchase energy whether used or not.

Utility Costs
- Physical Plant
- Transmission
- Substations
- Distribution Systems
- Meters
- Admin
- Energy
  — O&M
  — Fuel costs
- Interest on Debt
- Profit

Customer Rate Schedule Items
- Admin/Cust Charge
- Energy Charge
  — Energy in kWh
    - On-peak
    - Off-peak
  — Transmission (may be separate)
  — Capacity (may be separate)
- Fuel Cost Adjustment
- Demand Charge (may also be part of energy charge...)
- Demand Ratchet
- Power Factor
- Franchise Payment
- Sales Tax

Ratchet Clause: Demand is billed at a percentage (Usually > 50%) of the largest kW demand over the past 11 months, or the current month's demand, WHICHEVER IS GREATER.

Load Strategies
Off-peak Air Conditioning (OPAC) OPERATING STRATEGIES
  Load Leveling—Partial Load Shifting
- Partial shifting of AC load to off-peak hours
- Chiller runs at constant load or near constant load for 24 hours per day

— Very cost effective for new construction
— Less costly to purchase
— Less space needed
— But ~ less savings

Full Shift Strategy—operate at peak load hours only.

PUMP AND FAN LAWs

See Chapter 6

VARIABLE SPEED DRIVES

See Chapter 6

POWER QUALITY AND HARMONICS

Power quality describes how well the voltage waveform maintains a pure sinusoid. Equipment with switching power supplies such as computers, electronic ballasts, and variable speed drives will corrupt the waveform by introducing waveforms at various frequencies on top of the bus voltage. The waveforms introduced exist at multiples of the fundamental electrical frequency called **harmonics**. The lower, odd-value frequencies cause the most problems. The 3rd(180hz), 5th(300hz) and 7th(420hz) harmonics are of most concern. 3rd harmonics will induce currents to flow in neutral conductors which are typically sized smaller than phase conductors and may cause them to fail. 5th and 7th harmonics can create torque pulsations in motors leading to premature bearing failure.

IEEE PQ STANDARD 519

The Institute of Electrical and Electronics Engineers (IEEE) established the IEEE 519-1998 **Power Quality Standard**. It refers to Total Harmonic Distortion (THD) which is the ratio of the sum of the RMS amplitudes of the measured voltages for the higher frequencies divided by the RMS amplitude of the voltage at the fundamental frequency. THD can be measured using a harmonic analyzer. The standard recommends that for low voltage systems (less than 69 kV), the total harmonic distortion (THD) should not exceed 5% for general systems or 3% for specialty applications such as hospitals.

GROUNDING

Approximately 80% of power quality problems in facilities are related to poor electrical grounding. To treat grounding issues, wiring and grounding connections should be cleaned and tightened. For typical delta-wye load centers, the electrical neutral should be bonded to the equipment ground in the enclosure. Multiple ground loops should be eliminated.

LOAD FACTORS

The **load factor** is an estimate of the average percentage of full load operation for a facility or piece of equipment.

% Load Factor = Actual kWh consumed / (Peak kW x time) * 100

Example: If the peak kW for a given month at a facility was 750 kW and its measured energy use was 360,000 kWh, what is the Load Factor? Note that you must calculate the number of hours in a given month. If we assume 30 days X 24 hours/day, we get 720 hours.

%LF = (360,000 kWh)*100/(750 kW)*(720 hours) = 67 percent

THREE-PHASE SYSTEMS

Summary
- The **conductors** connected to the three points of a three-phase source or load **are called lines**.
- The three **components** comprising a three-phase source or load are

called **phases**.

- **Line voltage is the voltage measured between any two lines** in a three-phase circuit.
- **Phase voltage is the voltage measured across a single component** in a three-phase source or load. Typically measured as voltage difference between one line and the neutral connection.
- **Line current is the current through any one line between a three-phase source and load.**
- **Phase current is the current through any one component** comprising a three-phase source or load.
- **In balanced "Y" circuits,** line voltage is equal to phase voltage times the square root of 3, while **line current is equal to phase current.**

For "Y" circuits:

$$V_{line} = \sqrt{3}\ V_{phase}$$

$$I_{line} = I_{phase}$$

- **In balanced Δ (delta) circuits, line voltage is equal to phase voltage,** while line current is equal to phase current times the square root of 3.

For Δ (delta) circuits:

$$V_{line} = V_{phase}$$

$$I_{line} = \sqrt{3}\ I_{phase}$$

- Δ-connected three-phase voltage sources give greater reliability in the event of winding failure than Y-connected sources.
- Y-connected sources can deliver the same amount of power with less line current than Δ-connected sources.

A three-phase power system that connects three voltage sources together and uses a common connection point joining one side of each source is known as the "Y" circuit (Figure 4-2).

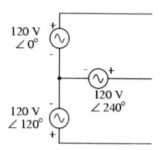

Figure 4-2. Three-phase "Y" circuit.

A "Y" circuit using coils for each leg is shown in Figure 4-3.

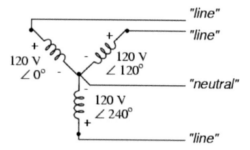

Figure 4-3. Three-phase, four-wire "Y" circuit.

A second three-phase configuration known as a "Delta" circuit is shown in Figure 4-4.

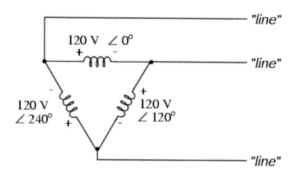

Figure 4-4. Three-phase Δ circuit.

SAMPLE PROBLEMS

1. A single-phase, 220-volt electrical load draws 5 kW and 27 amperes.
    What is its operating power factor?
    a. 49%
    b. 63%
    c. 75%
    d. 84%
    e. Can't determine from the information given

2. Why is a 277-volt lighting system used more often in a commercial
    building than a 120-volt system?
    a. Higher voltage results in greater lumen output
    b. Lower current resulting in lower power losses
    c. 277 equipment is more economical to purchase
    d. Power factor penalties are reduced with the higher voltage
    e. a and b

3. A utility penalizes customers for every kVAR in excess of 75% of
    the kW load. The minimum power factor that will entail no penal-
    ty is:
    a. 60%
    b. 70%
    c. 75%
    d. 80%

4. An electrical system consists of 40kW purely resistive load in par-
    allel with a 60kW partially resistive load with a power factor of
    80%. How many kVARs of capacitance are required to correct the
    power factor of the total load to 95%?
    a. 5.7 kVARs
    b. 12.7 kVARs
    c. 21.3 kVARs
    d. 45 kVARs
    e. 75 kVARs

5. The cause of most power quality problems in a typical facility is:
    a. electronic lighting ballasts
    b. rapid switching of SCRs

    c.   improper or inadequate grounding
    d.   telephony equipment
    e.   The electric utility

SOLUTIONS TO SAMPLE PROBLEMS

    1.   d
    2.   d
    3.   d
    4.   b
    5.   c

Chapter 5

Heating, Ventilation, and Air Conditioning (HVAC) Systems

INTRODUCTION

In most commercial buildings, the heating, ventilation, and air conditioning systems use a significant amount of energy. In northern climates, the heating loads are substantial and in southern climates the consumption is driven by air conditioning. If the facility has industrial refrigeration on site, that can consume more than 50% of the electricity at the facility. The exam topics are:

HVAC Topics

Affinity Laws	Performance (COP, EER, kW/ton)
Psychrometric Chart	HVAC Economizers
HVAC Equipment Types	Air Distribution Systems (Reheat, VAV)
Degree Days	Chillers
Heat Transfer	Energy Consumption Estimates
Vapor Compression Cycle	Absorption Cycle
Cooling Towers	Air and Water Based Heat Flow
ASHRAE Ventilation Standard	Demand Control Ventilation

HEAT TRANSFER DEFINITIONS

- **Conduction**—Transfer from hot side to cooler side through a medium
- **Convection**—Heat transferred between a moving liquid or gas and some conducting surface
- **Radiation**—Heat moves in waves (no medium)

AFFINITY LAWS

- **1st Law**: $CFM_2 = CFM_1*(RPM_2/RPM_1)$
  (Air flow is directly proportional to fan speed)
- **2nd Law**: $SP_2 = SP_1*(RPM2/RPM_1)^2$
  (Static pressure is a squared function related to fan speed)
- **3rd Law**: $HP_2 = HP_1 * (RPM_2/RPM_1)^3$
  (Fan horsepower is a cubed function related to fan speed)

PSYCHOMETRIC CHART

The **psychometric chart** is a graphical representation of the steam tables. Figure 5-1 represents the ASHRAE psychometric chart. The horizontal axis is temperature. The vertical axis is humidity ratio. The diagonal lines from lower right to upper left are enthalpy. The curved lines from lower left to upper right are constant humidity. The shaded area is the target operation zone for our HVAC system.

HVAC EQUIPMENT TYPES

- **Single zone**, which is used in one conditioned area, or with a group of areas with similar heat and cooling needs.
- **Multizone** systems use cooled and heated air that is mixed under the control of thermostats located at the individual zones.
  — **Dual duct systems** are of the multizone type with the air mixing done at the individual zone locations.
  — **Variable Air Volume** systems deliver varying amount of conditioned air at a constant temperature as required.
  — A **unit heater** has a fan and a heating coil but no ducting.
  — **Perimeter radiation** has heaters with no fans.
  — **Hot water converters** use a heat exchanger consisting of concentric pipes containing steam or hot water circulated around the conditioned air.

Constant Air Volume (CAV)

In CAV air handling system, the ventilation device is operated at a constant speed, which produces a constant flow of conditioned air.

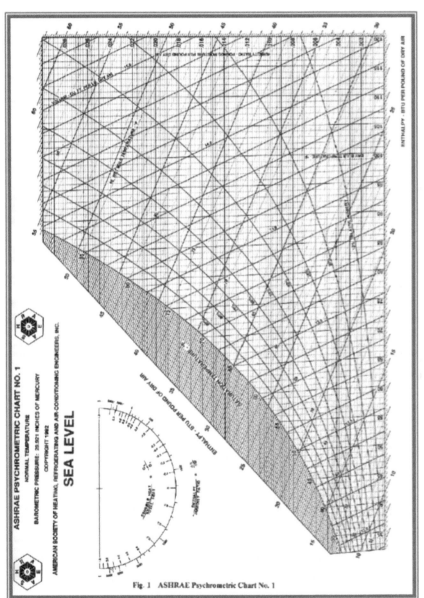

Figure 5-1. ASHRAE Psychometric Chart

Various control devices such as valves and dampers are operated to maintain the temperature. The **efficiency is limited**.

Variable Air Volume (VAV) Unit

A VAV unit offers variable speed control that modifies the volume of conditioned air that is delivered—VAV *controls temperatures by controlling air movement*. Usually the control is by *VFDs* for the fans but it may also be by mechanical adjustment. VAV systems are *inherently more efficient than CAV* as they only supply the air required to provide the required comfort. Power requirements relate to the cube of air movement; thus, less air volume means less electrical load.

> **Installing a fan inlet damper system** is a good strategy for conserving energy in a VAV system.

> Another possibility would be to **install static pressure controls** so that the pressure bypass dampers can be regulated more effectively.

HVAC deck systems use hot and cold decks to manage the intake of outside air and the output of indoor, conditioned air. For large HVAC systems, hot and cold decks are an efficient way for handling and conditioning hot and cold air according to the needs of each building zone. Generally:

• Air traveling over cooling coils is pushed into one duct holding cool air.

• Air traveling over heating coils is pushed into another duct storing hot air for ventilation.

A mixture of cool and hot air can also be supplied as the hot deck damper opens and the cold deck damper closes.

Terminal Reheat Air System (single zone system) supplies a constant volume of air, at a constant temperature, through a **single duct run**. This air handling fan supplies air through a run of duct and **provides air to heating coils. Heating coils cannot be modified** in a reheat system.
• Operates when the cooling load is less than maximum.

- Uses more energy, though provides a high degree of temperature and humidity control for a single zone.

The air temperature is selected low enough to cool the highest heat gain area, usually at 55 to 60 deg F. The thermostat-controlled reheat coils heat the air delivered to areas that do not require the full cooling capacity of the air.

Room thermostats modulate the reheat coil control valves to control the volume of heating water passing through the reheat coil to regulate the temperature of the air entering the area.

Increasing the air volume of single-zone units would reduce the energy efficiency of a terminal reheat HVAC system. Instead, the air volume of single-zone units should be decreased to promote efficiency.

Remotely Mixing Air

Induction systems—conditioned air is mixed with recirculated room air at individual control boxes. The primary air is delivered at high pressure and the addition of room air is induced by the flow of the cooled air from its nozzles.

A fan coil system *conditions all of the* **return air** in a small terminal called a fan coil unit (FCU). The FCU contains only a blower and a heating and cooling coil.

A **unit ventilator** conditions return and outside air, which is mixed as required. It circulates, heats, cools, and filters the air while allowing outside air in through a controlled damper.

DEGREE DAYS

Heating degree days and **cooling degree days** are indicators of energy consumption for indoor spaces. The basic assumption is that if the outside temperature is less than 65 degrees, a space must be heated assuming that we wish the indoor temperature to be 70 degrees. It is assumed that there is 5 degrees of heat internally from people and equipment.

Example

For three days, the outside temperature averages 50°F. the number of Heating Degree Days for that three day period is:

HDD = (65^{0}F – 50^{0}F) * 3 days = 45 degree days.

HEAT TRANSFER AND HEAT EXCHANGER TYPES

General Concepts

If exhaust gases are cooled below the dew point temperature, the water vapor in the gas will condense and deposit corrosive substances on the heat exchanger surface. Heat exchangers designed from low-cost materials will quickly fail due to chemical attack. Therefore, heat exchangers are generally designed to maintain exhaust temperatures above the condensation point.

HEAT EXCHANGER TYPES

A **metallic radiation recuperator** is the simplest type of heat exchanger. It comprises two metal tubes, one inside the other.

A **ceramic tube recuperator** is similar, but allows operation at much higher temperatures.

A **convective recuperator** consists of several parallel tubes through which hot gases are carried, as combustion air flows in the same direction along the outside.

A **combined radiation and convective recuperator provides the maximum effectiveness of heat transfer**.

A **double pipe heat exchanger** is the simplest exchanger used in industries.

Drawbacks: Low efficiency and they take up a large space.

Shell and tube heat exchangers consist of a series of tubes.

One set of tubes contains the fluid that must be either heated or cooled. The second fluid runs over the tubes (**in the shell**) that are being heated or cooled to either provide or absorb heat as required.

A set of tubes is called the **tube bundle** and can be made up of several types of tubes: **plain, longitudinally finned**, etc.

Shell and tube heat exchangers are typically used for high- pressure applications (with pressures greater than 30 bar (435+ psig) and temperatures greater than 260°C). This is because the shell and tube heat exchangers are robust due to their shape.

Several thermal design features must be considered when designing the tubes in the shell and tube heat exchangers:

- **Tube diameter**: Using a small tube diameter makes the heat exchanger both economical and compact. However, it is more likely for the heat exchanger to foul up faster and the small size makes mechanical cleaning of the fouling difficult.

- **Tube thickness**: The tube wall thickness is usually determined to ensure:
  — Enough room for corrosion
  — Flow-induced vibration has resistance
  — Axial strength
  — Availability of spare parts
  — Hoop strength (to withstand internal tube pressure)
  — Buckling strength (to withstand overpressure in the shell)

- **Tube length**: heat exchangers are usually cheaper when they have a smaller shell diameter and a long tube length.

- **Tube pitch**: when designing the tubes, it is practical to ensure that the tube pitch (i.e., the centre-centre distance of adjoining tubes) is not less than 1.25 times the tubes' outside diameter. A larger tube pitch leads to a larger overall shell diameter, which leads to a more expensive heat exchanger.

- **Tube corrugation**: this type of tube, mainly used for inner tubes, increases the turbulence of the fluids and the effect improves heat transfer performance.

- **Tube Layout**: refers to how tubes are positioned within the shell. There are four main types of tube layout, which are, triangular (30°), rotated triangular (60°), square (90°) and rotated square (45°). The triangular patterns are employed to give greater heat transfer as they force the fluid to flow in a more turbulent fashion around the piping. Square patterns are employed where high fouling is experienced and cleaning is more regular.

- **Baffle Design**: baffles are used in shell and tube heat exchangers to direct fluid across the tube bundle, run perpendicularly to the shell and hold the tube bundle, preventing sagging over a long length.
  - **Segmental baffle**—The semicircular segmental baffles are oriented at 180 degrees to the adjacent baffles forcing the fluid to flow upward and downwards between the tube bundle.
  - Baffles must be spaced with consideration for the conversion of pressure drop and heat transfer.
  - For thermo economic optimization it is suggested that the baffles be *spaced no closer than 20% of the shell's inner diameter*.
  - Having baffles spaced too closely causes a greater pressure drop because of flow redirection, but also important to ensure the baffles are spaced close enough that the tubes do not sag.
  - **Disc and donut baffle**, which consists of two concentric baffles. An outer, wider baffle looks like a donut, whilst the inner baffle is shaped like a disk. This type of baffle forces the fluid to pass around each side of the disk then through the donut baffle generating a different type of fluid flow.

Plate Heat Exchanger

These consist of multiple, thin, slightly separated plates that have **very large surface areas and fluid flow passages for heat transfer.** This stacked-plate arrangement can be more effective, in a given space, than the shell and tube heat exchanger.

While being less effective than rotary type systems, **fixed plate heat exchangers have no moving parts**. Plates consist of alternating layers of plates that are separated and sealed. Typical flow is cross current and since the majority of plates are solid and non permeable, sensible only transfer is the result.

The tempering of incoming fresh air is done by a heat or energy recovery core. **Humidity levels are adjusted through the transferring of water vapor**. This is **done with a rotating wheel** either containing a desiccant material or permeable plates. In this case, the core is made of aluminum or plastic plates.

- **Advances in gasket and brazing technology** have made the plate-type heat exchanger increasingly practical.

— In HVAC applications, **large heat exchangers** of this type are called **plate-and-frame**; when used in open loops, these heat exchangers are normally of the gasket type to allow periodic disassembly, cleaning, and inspection.

— There are many types of **permanently bonded plate heat exchangers**, such as dip-brazed, vacuum-brazed, and welded plate varieties, and they are often specified for closed-loop applications such as refrigeration.

— Plate heat exchangers also differ in the types of plates that are used, and in the configurations of those plates. Some plates may be stamped with "chevron," dimpled, or other patterns, where others may have machined fins and/or grooves.

Plate and Shell Heat Exchanger

• **Combines plate heat exchanger with shell and tube heat exchanger technologies.**

• The heart of the heat exchanger contains a fully welded circular plate pack made by pressing and cutting round plates and welding them together.

• Plate and shell technology offers **high heat transfer, high pressure, high operating temperature, compact size, low fouling and close approach temperature.** In particular, it does completely **without gaskets**, which provides security against leakage at high pressures and temperatures.

Adiabatic Wheel Heat Exchanger

Uses an **intermediate fluid or solid store to hold heat**, which is then moved to the other side of the heat exchanger to be released. Two examples:

• Adiabatic wheels, which consist of a large wheel with fine threads rotating through the hot and cold fluids

• Fluid heat exchangers.

Plate Fin Heat Exchanger

Uses "sandwiched" passages containing fins to increase the effectiveness of the unit. The designs include crossflow and counterflow coupled with various fin configurations such as straight fins, offset fins and wavy fins.

Plate and fin heat exchangers are usually made of **aluminum** alloys, which provide **high heat transfer efficiency.** The material enables the system to **operate at a lower temperature** and **reduce the weight of the equipment.**

Advantages
• High heat transfer efficiency especially in gas treatment (natural gas, helium and oxygen liquefaction plants, air separation plants and transport industries such as motor and aircraft engines).
• Larger heat transfer area.
• Approximately 5 times lighter in weight than that of shell and tube heat exchanger.
• Able to withstand high pressure.

Disadvantages
• Might cause clogging as the pathways are very narrow.
• Difficult to clean the pathways.
• Aluminum alloys are susceptible to mercury liquid embrittlement failure.

Pillow Plate Heat Exchanger

Commonly used in the dairy industry for cooling milk in **large direct-expansion stainless steel bulk tanks.** The pillow plate allows for **cooling across nearly the entire surface area of the tank, without gaps** that would occur between pipes welded to the exterior of the tank.

The pillow plate is constructed using a thin plate welded in a regular pattern of dots or with a serpentine pattern of weld lines. After welding the enclosed space is pressurized with sufficient force to cause the thin metal to bulge out around the welds, providing a space for heat exchanger liquids to flow, and creating a characteristic appearance of a swelled pillow formed out of metal.

Fluid Heat Exchangers

This is a heat exchanger with a gas passing upwards through a shower of fluid (often water). This is commonly used for cooling gases while also removing certain impurities, thus solving two problems at once.

Dynamic Scraped Surface Heat Exchanger

Another type of heat exchanger is called "(Dynamic) Scraped Surface Heat Exchanger." This is mainly used for **heating or cooling with high- viscosity products, crystallization processes, evaporation and high-fouling applications**. *Long running times* are achieved due to the continuous scraping of the surface, thus *avoiding fouling and achieving a sustainable heat transfer* rate during the process.

Phase-change Heat Exchangers

In addition to heating up or cooling down fluids in just a single phase, heat exchangers can be used either:

- To heat a liquid to evaporate (or boil) it
- Used as condensers to cool a vapor and condense it to a liquid.

In chemical plants and refineries, **reboilers** used to heat incoming feed for distillation towers are often heat exchangers. Distillation set-ups typically use condensers to condense distillate vapors back into liquid.

Waste Heat Recovery Units

A Waste Heat Recovery Unit (WHRU) is a heat exchanger that **recovers heat from a hot gas stream while transferring it to a working medium, typically water or oils**.

The hot gas stream can be the exhaust gas from a gas turbine or a diesel engine or a waste gas from industry or refinery.

Big systems with high volume and temperature gas stream, typical on industry, can benefit from Steam Rankine Cycle (SRC) in a WHRU.

The recovery of heat from low temperature systems requires more efficient working fluids than steam.

An Organic Rankine Cycle (ORC) WHRU can be more efficient at low temperature range using refrigerant that boils at lower temperatures then water. Typical organic refrigerants are Ammonia, Pentafluoropropane(R-245fa and R-245ca), and Toluene.

Challenges to Recovering Low-Temperature Waste Heat
Low temperature heat recovery faces at least three challenges:
- **Corrosion of the heat exchanger surface**: As water vapor contained in the exhaust gas cools, some of it will condense and deposit corrosive solids and liquids on the heat exchange surface. The heat exchanger must be designed to withstand exposure to these corrosive deposits.

- **Large heat exchange surfaces required for heat transfer**: Heat transfer rates are a function of the thermal conductivity of the heat exchange material, the temperature difference between the two fluid streams, and the surface area of the heat exchanger. Since low-temperature waste heat will involve a smaller temperature gradient between two fluid streams, larger surface areas are required for heat transfer. This limits the economics of heat exchangers.

- **Finding a use for low-temperature heat**: Recovering heat in the low-temperature range will only make sense if the plant has a use for low-temperature heat. Potential end uses include domestic hot water, space heating, and low-temperature process heating.

Technologies are available that can cool gases below dew point temperatures to recover low-temperature waste heat. Examples are:
- Deep economizers
- Indirect contact condensation recovery
- Direct contact condensation recovery
- Recently developed transport membrane condensers

Commercialization has been limited due to high costs and because facilities lack an end use for the recovered heat. When facilities lack an end-use for waste heat, some have found other means for recovery, including heat pumps and low-temperature power generation. These technologies are also frequently limited by economic constraints.

Direct contact heat exchangers

Direct contact heat exchangers involve heat transfer between hot and cold streams of two phases in the absence of a separating wall. Thus such heat exchangers can be classified as:

• Gas-liquid
• Immiscible liquid-liquid
• Solid-liquid or solid-gas

Such types of heat exchangers are used predominantly in air conditioning, humidification, industrial hot water heating, water cooling and condensing plants.

HVAC Air Coils

One of the widest uses of heat exchangers is for air conditioning of buildings and vehicles. This class of heat exchangers is **commonly called air coils, or just coils** due to their often serpentine internal tubing.

Liquid-to-air, or air-to-liquid HVAC coils are typically of modified crossflow arrangement. In vehicles, heat coils are often called heater cores.

On **the liquid side** of these heat exchangers, the common fluids are water, a water-glycol solution, steam, or a refrigerant.

• For heating coils, hot water and steam are the most common, and this heated fluid is supplied by boilers, for example.
• For cooling coils, chilled water and refrigerant are most common. When a refrigerant is used, the cooling coil is the evaporator in the vapor-compression refrigeration cycle, commonly called DX coils. Some DX coils are "microchannel" type.

On **the air side** of HVAC coils a significant difference exists between those used for heating, and those for cooling.

• Heating coils need not consider moisture condensation on their air side.
• Cooling coils must be adequately designed and selected to handle their particular latent (moisture) as well as the sensible (cooling) loads.
• Water that is removed is called condensate.

Spiral Heat Exchangers

A spiral heat exchanger (SHE) may refer to a helical (coiled) tube configuration; more generally, the term refers to a **pair of flat surfaces** that are coiled to form the two channels in a counter-flow arrangement.

The main **advantage** of the **Spiral Heat Exchanger** is its **highly efficient use of space.** This attribute is often leveraged and partially reallocated to gain other improvements in performance, according to well known tradeoffs in heat exchanger design (e.g., capital cost vs. operating cost).

- A compact SHE may be used to have a smaller footprint and thus lower all-around capital costs

- Over-sized SHE may be used to have less pressure drop, less pumping energy, higher thermal efficiency, and lower energy costs.

- Self cleaning: SHEs are often used in the heating of fluids that contain solids and thus tend to foul the inside of the heat exchanger. The low pressure drop lets the SHE handle fouling more easily.

Self-cleaning water filters are used to keep the system clean and running without the need to shut down or replace cartridges and bags.

There are three main types of flows in a spiral heat exchanger:

1. **Counter-current Flow:** Fluids flow in opposite directions.

2. **Spiral Flow/Cross Flow:** One fluid is in spiral flow and the other in a cross flow.

3. **Distributed Vapor/Spiral flow:** This **design is that of a condenser, usually mounted vertically.**

The SHE is good for **applications such as pasteurization, digester heating, heat recovery, pre-heating (see:** *recuperator*)**, and effluent cooling.** For sludge treatment, SHEs are generally smaller than other types of heat exchangers.

Heat Exchanger Selection Criterion
- Cost
- High/low pressure limits
- Thermal performance
- Temperature ranges
- Product mix (liquid/liquid, particulates or high—solids liquid)
- Pressure drops across the exchanger

- Fluid flow capacity
- Cleanability, maintenance and repair
- Materials required for construction
- Ability and ease of future expansion
- Material selection, such as copper, aluminum, carbon steel, stainless steel, nickel alloys, and titanium.

Typically in the manufacturing industry, several differing types of heat exchangers are used for just the one process or system to derive the final product. Example:
- Kettle (reboiler) HX for pre-heating
- Double pipe HX for the 'carrier' fluid
- Plate and frame HX for final cooling

Fouling occurs when impurities deposit on the heat exchange surface. Impurities can decrease heat transfer effectiveness significantly over time and are caused by:
- Low wall shear stress
- Low fluid velocities
- High fluid velocities
- Reaction product solid precipitation

Types of Fouling
- **Crude Oil Exchanger Fouling**. A series of shell and tube heat exchangers typically exchange heat between crude oil and other oil streams to heat the crude to 260°C prior to heating in a furnace. Fouling occurs on the crude side of these exchangers.
- **Cooling Water Fouling**. Cooling water typically has a high total dissolved solids content and suspended colloidal solids. To prevent fouling, designers typically ensure that cooling water velocity is greater than 0.9 m/s and bulk fluid temperature is maintained less than 60°C.

Maintenance
- **Plate** heat exchangers must be **disassembled and cleaned periodically**.
- **Tubular heat exchangers** can be cleaned by such methods as acid cleaning, sandblasting, high-pressure water jet, bullet cleaning, or drill rods.

- In large-scale cooling water systems for heat exchangers, **water treatment such as purification, addition of chemicals, and testing, is used to minimize fouling of the heat exchange equipment**. In steam systems for power plants, etc. to minimize fouling and corrosion of the heat exchange and other equipment.
- A variety of companies have started using water borne oscillations technology to prevent **biofouling**. Without the use of chemicals, this type of technology has helped in providing a low-pressure drop in heat exchangers.

VAPOR COMPRESSION CYCLE

The **vapor compression cycle** is viewed as consisting of four parts. There is a compressor, a condenser, an expansion valve and an evaporator. A refrigerant circulates within the loop. Assume that the refrigerant is in the liquid state with some vapor and is at low pressure as it enters the evaporator (labeled as step 4). If warm indoor air flows over the evaporator, the liquid refrigerant will absorb the heat and began to boil. The vapor that is created is pressurized by the compressor (labeled as step 1). This action elevates both the vapor pressure and temperature. The now hot vapor at high pressure enters the condenser (labeled as step 2). The heat that has been absorbed is released to the cooler air outside that flows over the condenser, and the vapor mostly returns to liquid form. This liquid then passes to the expansion valve (labeled as step 3). This allows the pressure of the liquid/vapor mix to lower. During expansion, the pressure of the refrigerant drops, then moves to the evaporator and the cycle repeats.

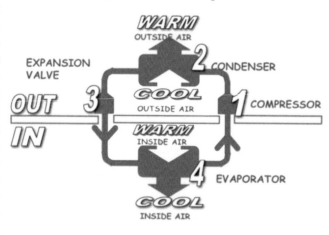

COOLING TOWERS

A cooling tower is a heat rejection device. **Over 1,000 Btu of heat can be released to the atmosphere for each pound of water evaporated.** Warm water is pumped outside to a cooling tower. It is then sprayed over large surfaces that are exposed to moving air driven by fans. The water is exposed to the cooler outside temperature and the moving air. A small portion of the water being cooled is evaporated. The water moves through the surface (known as the fill) and moves down via gravity. At the bottom of the tower the cooler water is gathered and returned to the building. Note that an indirect, or closed-circuit cooling tower involves no direct contact of the air and the water.

ASHRAE VENTILATION STANDARD

ASHRAE Standard 62 prescribes a ventilation standard of **15 cubic feet of outside air per building occupant**. This level may be ensured by controlling the indoor CO_2 content. The ventilation demand in each zone can be determined by remote CO_2 sensors in a similar manner as a thermostat that regulates the degree of cooling or heating supplied.

HVAC PERFORMANCE MEASURES

There are three primary HVAC performance measures that a CEM should be familiar with. The Energy Efficiency Ratio (EER) is the ratio of cooling output in Btu to the energy input in Watt-hours. It can also be expressed as the ratio of cooling input in Btu/hour divided by the Watts of electric input.

The **Seasonal Energy Efficiency Ratio (SEER)** is related to EER but is typically higher than EER because it accounts for typical variation in outdoor temperature. One ton of air conditioning (A/C) = 12,000 Btu/hr. A ton is a measure of A/C power, and is used when sizing systems, or when determining electrical demand.

Energy Efficiency Ratio (EER)

$$EER = \frac{\text{Btu of cooling output}}{\text{Wh of electric input}}$$

$$= \frac{\text{Btu/hr of cooling output}}{\text{W of electric power input}}$$

Coefficient of Performance (COP)

$$\text{COP} = \frac{\text{Energy or heat output (total)}}{\text{Energy or heat input (external only)}}$$

$$= \text{EER}/3.412 \text{ Btu/Wh}$$

Conversion between HVAC Measures

$$\frac{\text{kWin}}{\text{ton}} = \frac{12}{\text{EER}} = \frac{3.517}{\text{COP}}$$

HVAC ECONOMIZERS

Economizers in HVAC systems **facilitate the use of outside air as a cooling source when it is appropriate.** Considerations:

- The **air must be cold enough** and its **water content must also be low enough.**
- It **requires an additional fan** to move the outside air and equipment to sense the changeover point and control the flow.
- **Enthalpy controllers** use the air temperature and the wet bulb temperature or relative humidity to sense this point so that the total heat contents of the air sources are considered.
- In very dry climates, a simple temperature sensor may perform as well.

AIR DISTRIBUTION SYSTEMS (REHEAT AND VAV)

Air Distribution Approaches:
Single duct with single temperature zone
Single duct with reheat in terminal box
Single duct VAV system is the most common in commercial buildings
Dual duct system
Fan coil system

Temperature Control Strategies:
   CAV—Constant Air Volume—vary supply air temperature
   VAV—Variable Air Volume—vary flow while keeping temperature
      constant
   VAV-Reheat—vary both temperature and flow and reheat in
      terminal box.

CHILLERS

A chiller system consists of two water loops. In the bottom loop, the chilled water captures heat from the space as air travels over the air handling unit and returns it to the chiller. In the upper loop, the heat evaporates into the atmosphere in the cooling tower and the water gets chilled again.

ENERGY CONSUMPTION ESTIMATES

A pound of air costs more to move than a pound of water and the heat capacity of air is ¼ that of water.

For humidification or dehumidification at least 1000 Btu of energy is required per pound of water.

One ton of A/C = 12,000 Btu/hr (power term)

One ton-hour of A/C = 12,000 Btu (energy term)

Energy Efficiency Ratio(EER) = Btu of cooling output/Wh of electric input

Seasonal Energy Efficiency Ratio(SEER) = Seasonally averaged EER

Coefficient of Performance (COP) = Heat Removal(Btu or Btu/hr)/Energy input (Btu or Btu/hr)

kW/ton = 12/EER = 3.517/COP

Heating Seasonal Performance Factor (HSPF) = Heat provided(Btu)/Energy input(Wh)

COP = HSPF/3.412 Btu/Wh

ABSORPTION CYCLE

Absorption chillers are used when a business has available low-cost hot water or steam. Absorption chilling is a three-step process:

1.   Evaporation—Heat is extracted from space when a liquid refrigerant evaporates in a low pressure environment.
2.   Absorption—The gas refrigerant is absorbed by a liquid salt solution.
3.   Regeneration—The liquid is heated and the refrigerant evaporates. The hot refrigerant in gas form goes through a heat exchanger allowing the heat to be removed to another space (i.e. outside). This causes the gas to condense back to liquid as it returns to the evaporator.

Absorption chillers use no CFCs and have low efficiency (COP = 0.6 – 1.2), however, inexpensive source energy makes the process cost effective.

AIR AND WATER BASED HEAT FLOW

Water-cooled chillers need more accessory equipment than air-cooled units but they are more efficient. Estimates:

•   Large centrifugal water-cooled chillers (app. 45kW/ton or EER=27)
•   Large air-cooled chillers (app. 0.9kW/ton or EER=13)

DEMAND CONTROL VENTILATION

Demand control ventilation is a means of ventilation control where the **quantity of outside air mixed into the HVAC system is variable and determined by demand rather than being at a preset value.** This procedure can save considerable energy by reducing the quantity of warmer outside air introduced when it is not needed. It is often implemented by monitoring the CO_2 in the return duct from a space and introducing outside air to ensure that the CO_2 levels are below threshold values for maximum occupancy. The **alternative is to use a fixed mixing ratio based on the design occupancy levels** for each zone in the building.

SAMPLE PROBLEMS

1. Outside air has a dry bulb temperature of 80°F, and an enthalpy of 35 Btu/lb. What is the dew point temperature of the air?
    a. About 80 degrees
    b. About 75 degrees
    c. About 72.5 degrees
    d. About 67.5 degrees
    e. About 64.5 degrees

2. Approximately how much does it cost to cool 1 million cubic feet of air at 95°F and 70% relative humidity to 70°F and 100% relative humidity? The AC COP is 2.7 and electricity costs 8 cents/kWh.
    a. About $10
    b. About $20
    c. About $30
    d. About $40
    e. About $50

3. Which of the following climates best supports the retrofit of heat pipes in an air-conditioning system?
    a. High humidity
    b. Low humidity
    c. Moderate humidity
    e. None of the above

4. In a Variable Air Volume (VAV) system, energy savings result from optimizing the temperature of a constant volume of supply air.
    a. True
    b. False

5. A steam-driven chiller uses 18 lbs of atmospheric steam per hour to produce one ton of cooling. Without using steam tables, its COP

is estimated to be about:

a.   0
b.   1.0
c.   0.67
d.   1.5
e.   None of the above

SOLUTIONS TO SAMPLE PROBLEMS

1.   d
2.   a
3.   a
4.   b
5.   c

Chapter 6

Motors and Drives

INTRODUCTION

A knowledge of motors, applications, operation, and energy consumption is standard subject matter for an energy manager. The fundamentals of motor types and the potential for energy cost measures will be discussed.

In larger buildings, including manufacturing, industrial plants, supermarkets, commercial office, and multi-family housing, equipment including compressors, conveyors, hoists, pumps, and hydraulics may comprise a significant energy load. The management of these energy expenses and methods to reduce both operational hours and peak loads are paramount to controlling costs and reducing carbon footprint.

Topic Areas

AC Induction Motors	Variable Frequency Drives
AC Synchronous Motors	Fan and Pump Laws
DC Motors	Variable Flow Systems
High Efficiency Motors	Motor Selection Criteria
Load Factor and Slip	New vs. Rewound Motors
Power Factor and Efficiency	Motor Management Software
Motor Speed Control	Power Factor Correction

MOTOR BASICS

Motors convert electrical energy to mechanical energy.

Electromagnetic Induction is used to generate mechanical energy, or motion, by applying a current to a conductor wound around a magnetic material.

Poles of two magnets placed opposite each other will generate a magnetic field that causes a third magnet placed in the field to rotate (Figure 6-1).

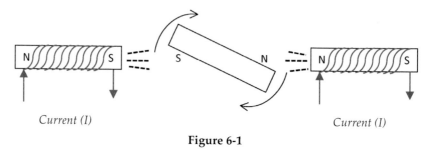

Current (I)   *Current (I)*

Figure 6-1

Motors may use one or more **pole pairs.**

The **conductor** wound around a magnet is referred to as the windings; windings are *stationary,* and thus also referred to as the **Stator.**

The **Rotor** is the rotating part of the motor.

Torque is the force required to produce rotation of a device.

Synchronous Speed is the theoretical speed of a motor based on the rotating magnetic field.

Alternating Current (AC) is a flow of electrical energy through a conductor that changes direction at a specified frequency in hertz (Hz). Common AC operating frequencies are 50Hz and 60Hz.

Synchronous speed depends upon the **AC frequency** and **number of poles** of the machine.

Direct Current (DC) is a flow of electrical energy through a conductor in one direction at a constant value.

TYPES OF MOTORS

AC Induction Motor (Asynchronous Motor)

An **induction motor** *always runs at a speed less than synchronous speed.*

There are basically **two types of induction motors** that depend upon the input supply:
• **single-phase induction motor** (not self-starting)
• **three-phase induction motor** (self-starting)

Rotor speed will depend upon the AC supply and is controlled by varying the input supply. This is the working principle of an induction motor of either type.

Single-phase Induction Motor
- Split phase induction motor
- Capacitor start induction motor
- Shaded pole induction motor

A single-phase induction motor **is not self-starting**: AC is applied and the resulting oppositely rotating magnetic fields do not produce a **torque** to turn the rotor. An external force must be applied to move the rotor and start motor operation.

Three-phase Induction Motor
- Squirrel cage induction motor
- Slip ring induction motor

- A three-phase induction motor **is self-starting**: AC is applied and the magnetic fields created by the three phases turns the rotor.

Advantages of induction motor:
- Efficiency up to 97%.
- The direction of rotation of induction motor can easily be changed by changing the sequence of three-phase AC supply.

Disadvantage of induction motor:
- The **speed** of the motor **varies with the motor load**.

AC Synchronous Motor
   Synchronous Motor—the speed of the rotor equals the speed of the rotating magnetic field, and therefore in sync with the AC supply frequency.
   Rated Speed—the actual speed the motor operates when fully loaded at the supplied rated (nameplate) voltage.

Synchronous Motor Features
- Inherently **not self-starting**: requires external means to bring motor speed close to its synchronous speed before fully synchronized.
- **Constant speed** irrespective of load condition.
- Operates under any electrical power factor and thus can be used to improve power factor (see *power factor* in this chapter).

Synchronous Motor Applications
- Lower operating speeds (around 500 rpm) and high power (35 kW to 2500 kW)—preferred to three-phase induction motor whose size, weight and cost is very high.
- Power factor improvement without a load—used where static capacitors are expensive.

Examples
- Reciprocating pump
- Compressor
- Rolling mills

DC Motor

A **DC Motor** is used to convert electrical energy to a mechanical energy by creating a magnetic field as in AC motors.

The DC Motor is capable of **maintaining the same speed under variable load**.

MOTOR EFFICIENCY AND OPERATION

Motor efficiency is the ratio between shaft output power and electrical input power.

Full Load RPM (FLRPM)—speed at which the motor will rotate at rated voltage and frequency during full force to produce rotation, called **torque**.

Motor Slip—percent difference between a motor's rated speed (FLRPM) and synchronous speed, an indication of how much the speed will vary.

Low slip motors (small difference between synchronous and actual speed) are **high efficiency motors**.

High slip motors (low efficiency) are used for applications where the load varies significantly.

Motor Efficiency Factors
Energy Savings Opportunities:
- Keep air filters clean
- Avoid running at partial load or unloaded

- Optimize by sequencing
- Use variable speed devices

Motor Load Factor

Motor Load Factor—the percentage of kW load of the motor's rated load (nameplate horsepower) times the motor efficiency:

The load factor is used to calculate motor load as follows:

$$\text{Motor Load (kW)} = \frac{\text{NPHP x (.746 kW/HP) x (Load Factor)}}{\text{Efficiency}}$$

Slip Method

A motor's speed and slip is proportional to its load. Slip can be used with measured synchronous speed to calculate the % motor load:

$$\% \text{ Motor Load} = \frac{\text{Time Slip}}{\text{Design Slip}} = \frac{(S_{\text{synchronous}} - S_{\text{measured}})}{(S_{\text{synchronous}} - S_{\text{Full Load}})}$$

Power Factor, Efficiency, and Correction

A motor or other magnetic device (magnetic ballast, for example) uses single-phase or three-phase power to perform work (compression, air flow, hoisting, etc.). Power is also consumed due to the magnetic induction of the motor, in which current through the windings is *out of phase* from the current used to perform work. The power consumed by magnetic induction is lost in the form of heat.

Real Power—power consumed that performs work.

Reactive Power—power consumed by magnetic induction

Total (Apparent) Power = Real Power + Reactive Power

Power Factor indicates motor efficiency as the ratio of real power to total power:

$$\text{Power Factor (pf)} = \frac{\text{Real Power}}{\text{Total Power}}$$

Power Factor is noted as a value between 0 and 1.

Examples

- A motor with a relatively high electric load consumes more real power, and thus has a higher power factor.

- A motor with a relatively low load (or no load) would consume very little real power, and thus has a low power factor; nearly all power consumed is reactive.

As discussed in Chapter 1, power factor is **corrected** by adding **capacitance** to a circuit to adjust the phase difference between real and reactive currents, and thus more real power is consumed to perform work.

Power factor can be calculated for single-phase and three-phase circuits.

Motor Speed Control

Speed of a motor is determined by the frequency of AC current (Hz) and the number of poles used.

- At zero load, the motor operates at (or very near) the synchronous speed.
- At full load, the motor operates at its rated speed = nameplate speed.

$$\text{Speed}_{synchronous} \text{ (rpm)} = 120 \times \text{Frequency (Hz)} / \# \text{ Poles}$$

# of Poles	Synchronous Speed (rpm)
2	3600
4	1800
6	1200
8	900

Note: the chart lists number of poles, NOT pole pairs.

Variable Frequency Drives

Variable Frequency Drive (VFD)—electronic device that varies the current frequency (Hz) to control the speed of a motor. A VFD reduces energy consumption by slowing a motor's rotation. Examples: reduce speed of fans, pumps, and compressors to meet required work demands in varying conditions.

VFDs are types of variable speed drives (VSDs).

Fan and Pump Laws

Affinity Laws—principles that govern the movement of air (gas) and water (liquid) (Fan/Pump Laws) at constant shaft speed.

Air

Law 1a. **Flow** is proportional (per foot, e.g.) to **shaft speed** N_x (cubic feet per minute, CFM)

$$\frac{\text{Flow } Q_1}{\text{Flow } Q_2} = \frac{\text{Speed } N_1}{\text{Speed } N_2}$$

Law 1b. **Pressure**, or "**Head**" (per in², e.g.) is proportional to the **square** of **shaft speed**:

$$\frac{\text{Pressure } H_1}{\text{Pressure } H_2} = \left(\frac{\text{Speed } N_1}{\text{Speed } N_1}\right)^2$$

Law 1c. **Power** (kW, e.g.) is proportional to the **cube** of **shaft speed**:

$$\frac{\text{Pressure } H_1}{\text{Pressure } H_2} = \left(\frac{\text{Speed } N_1}{\text{Speed } N_2}\right)^3$$

Water

Pump Impeller—a rotor used to increase the pressure and flow of a fluid.

Law 2a. **Flow** is proportional to the **impeller diameter** (inches, e.g.)

$$\frac{\text{Flow } Q_1}{\text{Flow } Q_2} = \frac{D_1}{D_2}$$

Law 2b. **Pressure** (Head) is proportional to the **square** of **impeller diameter**:

$$\frac{\text{Pressure } H_1}{\text{Pressure } H_2} = \left(\frac{D_1}{D_2}\right)^2$$

Law 2c. **Power** is proportional to the **cube** of **impeller diameter**:

$$\frac{\text{Power P}_1}{\text{Power P}_2} = \left(\frac{D_1}{D_2}\right)^3$$

Variable Flow Systems

Variable Flow Systems—systems that vary motor speed to regulate the flow of a liquid.

Example: a Variable Refrigerant Flow (VRF) system allows for energy savings when partial-load conditions are adequate to meet heat removal requirements.

Motor Selection Criteria

Motor Nameplate Terms—ratings provided for a motor which can be used to calculate motor load (power) in kW.

Full Load RPM (FLRPM)—speed at which the motor will rotate at rated voltage and frequency during full torque. This "full load" speed will normally vary between 87% and 99% of synchronous speed depending on design slip.

Insulation. Insulation is crucial in a motor to withstand the greatest temperature that occurs at the hottest point within the motor for as long as the temperature normally exists.

Time rating. Continuous duty will be shown as "CONT" on the nameplate.

Horsepower. Horsepower is determined by the output when the motor is loaded to **rated** torque at **rated** speed.

Locked Rotor Indicating Code Letter. When a motor is started, there is an 'inrush' of current; standardized and defined by a series of code letters which group motors based on the amount of inrush in terms of kilovolt amperes. The code letter defines low and high voltage inrush values on dual voltage motors. These values can be used for sizing starters, etc. (Figure 6-2).

Motor Service Factor (SF)—factor multiplied by horsepower to calculate the allowable horsepower loading of the motor.
• Provides a buffer between estimated horsepower requirement and actual horsepower operation.

Code	KVA/HP	Approx. Mid-Range Value
A	0.00-3.14	1.6
B	3.15-3.54	3.3
C	3.55-3.99	3.8
D	4.00-4.49	4.3
E	4.50-4.99	4.7
F	5.00-5.59	5.3
G	5.60-6.29	5.9
H	6.30-7.09	6.7
J	7.10-7.99	7.5
K	8.00-8.99	8.5
L	9.00-9.99	9.5
M	10.00-11.99	10.6
N	11.20-12.49	11.8
P	12.50-13.99	13.2
R	14.00-15.99	15.0

Figure 6-2. Locked Rotor Indicating Codes

- Allows for cooler winding temperatures at rated load.
- Protects against intermittent heat rises.
- Helps offset low or unbalanced line voltages.

Efficiency. Efficiency is the ratio of the power output divided by the power input. The efficiency is reduced by any form of heat, including friction, stator winding loss, rotor loss, core loss (hysteresis and eddy current), etc.

NEMA Design Letter—designation for specified locked rotor torque, breakdown torque, slip, starting current, or other values. NEMA design letters are A, B, C, and D.

- NEMA Design A motors have normal starting torques, but high starting currents. This is useful for applications with brief heavy overloads. Injection molding machines are a good application for this type of motor.
- NEMA Design B motors are the most common. They feature nor-

mal starting torque combined with a low starting current. These motors have sufficient locked rotor torques to start a wide variety of industrial applications.

- NEMA Design C motors have high starting torques with low starting currents. They are designed for starting heavy loads due to their high locked rotor torques and high full load slip.

- NEMA Design D motors have high starting torque and low starting current, however they feature high slip. This reduces power peaks in the event that peak power is encountered, motor slip will increase.

Enclosure Type. Open drip-proof (ODP), totally enclosed fan cooled (TEFC), explosion proof (EXP), totally enclosed non-vent (TENV), totally enclosed chemical duty, and totally enclosed wash down.

Manufacturer's Identification Numbers. The model, date, & serial number are supplied to aid in identification.

Bearing Part Numbers. The bearing part numbers are included if replacement bearings need to be obtained.

Connection Diagrams. This diagram is to aid a qualified electrician in the wiring of a motor.

MOTOR MAINTENANCE AND HARMONICS

New vs. Rewound Motors

Motor rewinding—repairs a failed motor
- Less expensive than motor replacement
- Damage resulting in a **1-2% loss in efficiency** is fairly common during the rewind process
- May increase efficiency if enough **space** is available within the frame and **adding more copper** for the windings

Maintenance Factors
Belts—Cogged v-belts and synchronous v-belts can increase efficiency 2-4% vs. V-belts that may slip

Lubrication—reduce friction losses, save 1-2%

Periodic Maintenance—check:
- Operating temperature
- Electrical
- Bearing condition by using industrial stethoscope to measure noise

Vibration Analysis to find:
- Bad bearings
- Bad gears
- Failing machine mounts

Harmonics

Harmonics are additional waves of energy created as a multiple of the generated wave frequency in a *non-linear load* such as a motor, variable speed drive, or fluorescent light. Harmonics can interfere with equipment operation.

IEEE-519 PQ (Power Quality) Standard—To minimize the impact of facility harmonic distortion on the utility power system and on neighboring facilities.

The standard provides recommended limits for total harmonic voltage and current distortion.

The pulsed current waveforms from VFDs contained not only the 60hz components and **harmonics**. Effects:

- High peak current
- Elevated true RMS current
- Lower total power factor

Current that flows back in multiples of the fundamental is called **harmonic current**.

The **third harmonic** is three times the fundamental of 60 Hz (180 Hz). The **fifth harmonic** current is at 300 Hz, and so on.

- **Third harmonic** causes overheating in neutral conductors and transformers
- **Fifth harmonic** can cause motor issues, such as overheating, abnormal noise and vibrations, and motor inefficiency.

Other typical non-linear loads added during energy upgrades include electronic ballasts, computers, programmable logic controllers (PLCs), and various components of building automation systems.

MOTOR MANAGEMENT SOFTWARE

Motor Load Calculation Method:
Computerized Modeling Techniques
    MotorMaster+—no cost software for motor selection, efficiency analysis, and economic analysis.

•   National Electrical Manufacturers Association (NEMA) software

•   Includes a catalog of more than 20,000 low-voltage induction motors

•   Includes motor inventory management tools, maintenance log tracking, energy accounting, and environmental reporting capabilities.

    Equivalent circuit model: a mathematical model of the motor is made that takes into account for friction losses, winding losses, stray load losses, etc.

    ORMEL 96—program developed by Department of Energy (DOE)'s Oakridge National Lab that uses an equivalent circuit technique to determine motor load.

•   Being implemented in the **Motor Master+** Software
•   Based on IEEE's standard 112

SAMPLE PROBLEMS

1.   A 100 HP motor is running at 75% loading on 480 volt 3 phase electric power and 90% efficiency. What amperage is it consuming? Power factor is 85%.
    a.   110 amps
    b.   88 amps

    c.   20.7 amps

    d.   75 amps

    e.   Not enough information

2.   A 100 HP induction motor is running at 80% load and 85% efficiency. A new 100 HP energy efficient motor can be purchased that will run at 90% efficiency (80% loading), what is the savings in input electric power?

    a.   70 kW

    b.   9.6 kW

    c.   3.9 kW

    d.   15.2 kW

    e.   1.1 kW

3.   Variable speed drive application for a fan with a 20 HP motor is presently supplying 2500 CFM running at 1750 RPM. If the fan is slowed to a new RPM of 1000, how much motor HP is needed now?

    a.   2.51 HP

    b.   3.73 HP

    c.   5.15 HP

    d.   7.32 HP

    e.   Not enough information

4.   For an AC induction motor with 3 pole pairs (6 poles), the actual supplied frequency is 40 Hz (cycles per second.) What is the synchronous speed of the motor at 40 Hz?

    a.   1800 RPM

    b.   1200 RPM

    c.   800 RPM

    d.   600 RPM

    e.   400 RPM

5.   A facility presently uses a water pump with a 100 HP 3-phase motor with an efficiency of 92%. The pump operates 20 hours per day, 365 days per year. The motor is fully loaded when the flow rate is maximum. However, the actual flow rate needed over time varies in the following manner:

            6:00 am-noon   50% of maximum

| Noon-midnight | 75% of maximum |
| Midnight-2 am | 100% of maximum |

A VFD is installed to control the pump. Approximately how many kWh/yr are needed now with the VFD? Assume the motor efficiency is constant, and there are no losses in the VFD.

a. 1 million kWh/yr
b. 590,000 kWh/yr
c. 230,000 kWh/yr
d. 150,000 kWh/yr
e. 60,000 kWh/yr

SOLUTIONS TO SAMPLE PROBLEMS

1. e
2. c
3. c
4. b
5. b

Chapter 7

Industrial Systems

INTRODUCTION

Many industrial processes use energy intensive equipment requiring proper selection, integration, and choice of fuel for operational efficiency and energy cost management.

The sections in this chapter discuss equipment and applications that are essentially used to alter temperatures and pressures of liquids and gases for a wide range of industrial applications including manufacturing, material processing, maintenance services, and refrigeration.

Topic Areas

Waste Heat Recovery	Boilers and Thermal Systems
Industrial Energy Management	Fuel Choices
Steam Systems	Steam Tables
Heat Exchangers	Compressors
Turbines	Pumps and Pumping Systems
Compressed Air Systems	Air Compressors
Air Compressor Controls	Air Leaks

WASTE HEAT RECOVERY

Heat that is lost to the atmosphere or other medium without performing useful work is referred to as waste heat. **Waste heat Recovery** is the process to capture lost heat to perform useful work.

A cost-benefit analysis can determine whether to recover waste heat based on factors including:
- Energy produced
- Fuel savings
- Increased thermal capacity
- Capital expense

Heat exchangers (also see Chapter 5) are used to recover waste heat, most commonly from combustion exhaust gases to combustion air entering a furnace. Preheated combustion air reduces the amount of required fuel energy and associated cost.

Heat exchanger equipment for pre-heating combustion air includes the following:

Recuperators—medium to high temperature applications
- Radiation recuperator—waste gases pass through an inner duct and transfer heat to an outer duct
- Convective recuperator—waste gases pass through small diameter tubes and transfer heat to air within a larger shell
- Combined radiation/convection recuperator—radiant section followed by a convective section

Regenerators
- Regenerator Furnace—high temperature applications with dirty exhausts
  — Two brick chambers constructed in checkerboard (cross-hatch) patterns
  — Hot combustion exhaust gas flows through one chamber and transfers heat to incoming combustion air in the other chamber
  — Combustion exhaust and incoming combustion air alternate between chambers about every 20 minutes
  — Large size; high capital costs are significantly greater than recuperators

- Rotary Regenerator/Heat Wheel—low and medium temperature applications
  — Heat is stored in a porous media
  — Hot and cold gas flows alternate between two parallel ducts
  — Rotating porous disc (heat wheel) resides within the two ducts and transfers heat from the hot gas duct to the cold gas duct
  — Applications generally restricted from high temperatures due to the thermal stress of duct materials
  — Can be designed to recover moisture as well as heat from clean gas streams

Passive Air Preheaters—low to medium temperature applications where gas streams are isolated to prevent cross-contamination.
- Platetype exchanger preheater—multiple parallel plates that create separate channels for hot and cold gas streams
  — Hot and cold flows alternate between the plates and allow significant areas for heat transfer.
  — Systems less susceptible to contamination compared to heat wheels, but more susceptible to fouling
- Heat pipe exchanger preheater—multiple pipes with sealed ends
  — Hot gases pass over one end of a heat pipe causing the fluid inside the pipe to evaporate
  — Hot vapor moves to other end of sealed pipe and condenses, transfers heat to the cold gas

Regenerative/Recuperative Burners—burners that incorporate regenerative or recuperative systems.
- Simple, compact design
- Higher energy efficiency vs. burners using ambient air
- Burner design includes heat exchange surfaces to capture exiting stack (exhaust) gas heat
- Lower cost, lower energy recovery than stand-alone units

Finned Tube Heat Exchangers/Economizers—low to medium temperature heat recovery from exhaust gases to heat liquids
- Round tube with attached fins maximize heat transfer surface area
- Hot gases flow across tubes transferring heat to liquid
- Feedwater preheating is common application (boiler economizer)

Waste Heat Boilers—water tube boilers that use exhaust gases to generate steam for process heating or for power generation.

BOILERS AND THERMAL SYSTEMS

Industrial **thermal systems** transfer heat to and/or from solids, liquids, or gases for an intended process. Thermal systems include heating and cooling plants, process steam generation, and furnaces, as examples.

Heat transfer occurs by one of three methods:

- **Conduction**: heat transfers from one location to another without use of the flow of a material.

- **Convection**: heat transfers from one location to another by the movement of a fluid (such as water or air, e.g.).

- **Radiation**: heat transfers by means of electromagnetic waves. All objects radiate energy via electromagnetic waves.

Industrial thermal systems include:

- Steam Boiler Systems
- Biomass Energy Systems
- Thermal Fluid Systems
- Electric Process Heaters
- Process Bath Heaters
- Direct Fired and Gas Fired Heaters
- Waste Heat Recovery

A **boiler** is a chamber that holds water to be heated to a desired temperature and pressure. Heated water in the form of steam is used as a medium to carry heat for industrial processes (see Chapter 15 for boiler steam distribution systems).

$$\text{Steam Boiler System Efficiency \%} = \frac{\text{Heat Output of Steam (Btus)}}{\text{Heat Input from Fuel (Btus)}} \times 100$$

In a fire-tube boiler:

- The burner combusts a fuel and transfers heat by *conduction* to the inside of the boiler tubes

- Heat within the tubes is transferred to the water on the outside tube surfaces in the boiler by conduction through tube walls

- The heated water on the outside tube surfaces transfers heat to water further away from the tube surfaces within the boiler by *convection* (motion of fluid).

STEAM SYSTEMS

Heat balance calculations are used to identify steam boiler system efficiencies and potential measures to improve performance, such as employing waste heat recovery, as an example.

In a heat balance calculation, based on the 1st law of thermodynamics in which energy can be transformed but cannot be created or destroyed:

Sum of Energy Inputs = Sum of Energy Outputs

Energy Inputs:
• Heat Gains
   — Fuel
   — Combustion air
   — Condensate return

Energy Outputs
• Steam
• Heat losses
   — Stack (flue) gases
   — Stack surface losses
   — Boiler blowdown
   — Boiler surface losses

STEAM TABLES

Enthalpy: measure of the following three forms of heat contained in a substance per mass unit
• **Sensible Heat**: heat content of steam due to temperature change
• **Latent Heat**: heat content of steam required to change state of the substance
• **Kinetic Energy**: energy due to the movement of particles (negligible for steam)

Enthalpy of steam is the heat content per mass (Btu/lb) for given pressures and temperatures. Enthalpy values are listed in a **steam table**. Steam table data are used for designing boiler system heat capacities and fuel requirements, cost estimates, steam leak costs, and other calculations.

Steam Table Examples

Sensible Temperature (°F)	Latent Pressure (psia)	Total Heat (Btu/lb)	Specific Heat (Btu/lb)	Heat (Btu/lb)	Volume (ft³/lb)
—	—	—	—	—	—
300	67.05	269.7	910.0	1179.7	6.466
310	77.67	280.0	902.5	1182.5	5.626
—	—	—	—	—	—
—	—	—	—	—	—

BOILER FUEL CHOICES

Boiler fuel choices are based on the boiler application, fuel availability, and economics of the fuel supply and delivery costs. In summary:

Coal
- Burner technology complicated but well-developed
- Sulfur Oxide (SO_x) and Nitrogen Oxide (NO_x) emissions, regulated by the EPA, and particulate matter (PM) emissions must be controlled
- Material handling costs are high (unloading, transport to the boiler, storage costs)

Fuel Oil
- Multiple classifications:
  - Number (No.) 1 Distillate (Kerosene, e.g.)
  - No. 2 Light Heating Oil
  - No. 4 Heating Oil (blend of No. 2 and No. 6)
  - No. 6 Heavy Heating Oil
- Burner technology well-developed
- Sulfur Oxide (SO_x) and Nitrogen Oxide (NO_x) emissions, regulated by the EPA, and particulate matter (PM) emissions must be controlled, especially for No. 6 heavy oil
- Delivery methods and storage practices are well established

Natural Gas
- Burner technology less complicated than coal, well-developed
- Sulfur Oxide (SO_x) and Nitrogen Oxide (NO_x) emissions are much lower than coal and oil
- Negligible particulate matter (PM) emissions
- Low cost transport through gas pipelines; end-use availability based on local delivery pipelines

Other Fuels
- Types
  — Biodiesel (produced or recovered waste vegetable oil)
  — Manufacturing waste
    a) Pulp mill inorganic chemicals
    b) Plywood production waste (hog fuel/wet bark)
    c) Sawdust
    d) Municipal garbage
    e) Food processing waste
  — Soft and hard woods
  — Wood pellets
- Issues
  — Combustion mix
  — Burner modifications
  — Residue materials and disposal
  — Transport to or within facility
  — Storage

Fuel heating values and energy densities are found in the Appendix.

INDUSTRIAL ENERGY MANAGEMENT

Industrial energy management is best managed by individuals that possess:
- Knowledge of energy systems and equipment elements in a facility
- Abilities to:
  — Direct staff
  — Prepare and present cost saving initiatives for upper management review

— Obtain the budgets necessary to continually improve energy efficiency in support of company business goals

Five key areas for industrial energy management are as follows:
1) Establish an Energy Management Team
   — Effective energy team leader
   — Energy staff members from multiple company departments
   — Technical analysis, cost saving estimates
   — Prepare/present proposals for funding to upper management/decision makers.
2) Obtain energy data to measure and document energy cost savings
   — Utility bill data
   Energy management system real-time data
   • By department
   • By process
   • By system
   • By equipment element
   • Weather adjusted
   — Energy consumption analysis software
3) Set business driven energy goals
   — Establish relevant, realistic, defensible goals
   — Use realistic metrics
   — Adhere to standards (ASHRAE, ISO 50001, e.g.)
4) Leverage resources to meet energy management goals
   — Energy management systems and software
   — Consultants and subject matter experts
   — Certified audit and retro-commissioning professionals
5) Communication to stakeholders
   — Energy studies
   — Recommendations and supporting data
   — Implementation planning
   — Measured savings
   — Green building certification achievements

TURBINES

Turbines are rotary engines that convert pressure and motion to rotational energy.

The turbine's **rotor** is the one primary moving part that includes a central shaft with blades to respond to the motion and pressure of the applied gas or liquid.

- **Impulse Turbines** respond to motion because they convert the fluid's impulse (mass and velocity) to torque.
- **Reaction Turbines** convert the fluid's pressure to rotary motion. The fluid at the exhaust side of the turbine contains either lower pressure or less velocity than at the input.

Turbines can be driven by gas, steam, or water.

PUMPS AND PUMPING SYSTEMS

Pumps are motor driven devices used to move liquids. The inlet of a pump is known as the suction port (suction side), and the outlet is known as the discharge port (discharge side).

- Centrifugal pumps use a rotating impeller to increase the pressure and flow rate of a fluid. Centrifugal pumps are the most common type of pump used to move liquids through a piping system.
- Positive displacement pumps force liquid to move by compressing a confined volume of liquid and forcing it into the discharge pipe.
  - Designed for constant discharge flow; a slight increase in internal leakage as pressure increases prevents a truly constant flow rate.
  - Some positive displacement pumps use an expanding cavity on the suction side and a decreasing cavity on the discharge side while maintaining a constant flow of volume.
  - Two types:
    - *Rotary pump*: uses one part or several parts to move the fluid in a circular fashion and discharges liquid with each revolution
    - *Reciprocating pump*: the fluid flows back and forth and released by the movement of a device, usually a piston or diaphragm

COMPRESSORS

Compressors are motor driven devices used to move gases (vapor). As with a pump, the inlet of a compressor is known as the suction port (suction side), and the outlet is known as the discharge port (discharge side).

Types

- **Reciprocating Compressor** (or piston compressor)
  — Positive-displacement compressor increases pressure of the gas by reducing volume
  — Uses a piston within a cylinder as the compressing and displacing element
  — Single-stage and two-stage (higher pressures) reciprocating compressors are available
  — Typically 1 to 50 HP
- **Centrifugal Compressor** (radial compressor)
  — Uses a **rotating impeller** to transfer energy to the gas
  — Dynamic displacement creates high pressure discharge by converting energy from rotating impeller momentum
  — Higher speed rotation required vs. other compressors
  — Designed for higher capacity
  — Continuous flow through the compressor
  — Oil-free compressor by design
- **Rotary Screw Compressor** (radial compressor)
  — Uses two meshing helical screws, known as rotors (male and female) to compress the gas.
  — Timing gears ensure that male and female rotors maintain precise alignment
  — Gas enters at the suction side, moves through screw threads as they rotate, forcing gas through the compressor to discharge port at higher pressure
  — High speeds must be used to minimize the ratio of leakage flow rate to effective flow rate

COMPRESSED AIR SYSTEMS

Compressed air systems are comprised of the following elements:
- Compressors
  — Positive displacement: increases pressure and reduces volume of air
  — Dynamic displacement: increases air pressure by converting

energy from rotating impeller momentum

- Air filters
  - — Removes particles, Particulates, and lubricants
  - — Upstream and/or downstream use
  - — Keep clean so as not to cause pressure drop/throttling
- Air dryers
  - — Refrigerated—most common—low initial and operating costs
  - — Regenerative Desiccant—use desiccant to dry—needs pre-filter (air filter)
  - — Heat of Compression—for lubricant free Rotary Screw Compressors
  - — Deliquescent Desiccant—dissolvable; regular replacement necessary, costs labor and material—needs pre-filter (air filter)
  - — Membrane type—for low volume applications
- Air distribution equipment
- Compressed air tanks
- Pressure/flow regulators
- Separators
- Condensate drains
- Supplementary equipment
  - — Optimizes compressed air for applications
  - — Storage
  - — Transport
  - — Flow and pressure regulation

Compressed Air Systems Energy Use and Efficiency—General Information

For every 1 psi drop from compressed air source, there is a .5% loss of energy. Example: if a filter clog causes a 4 psi drop, than 2% energy is lost due to this drop.

For every 1 psi reduction in pressure setting above 100 psi, there is a .5% energy saved. Example: if the air compressor pressure setting is reduced from 150 to 140 psi, there is a 5% energy savings.

AIR LEAKS

- Maintenance should include periodic leak inspection
- Leak(s) are more likely if compressor runs continuously

- Ultrasonic leak detection used in noisy environments; listen for leaks when nose levels permit

AIR COMPRESSOR CONTROLS

Compressed air system controls deliver compressed air supply for the given demand and have significant impact on overall system energy efficiency.

Key features of compressed air system controls:
- Designed to operate within a fixed pressure range known as the control range
- Delivers volume of air per system demand

Monitors system pressures to control compressor use:
- Decreases compressor output when the pressure reaches a predetermined level
- Increases compressor output when the pressure drops to a lower predetermined level
- Controls are specified for a given system by the type of compressors used and the facility's demand profile

Control strategies are used for both single compressor and multiple compressor systems.

Single Unit Control Schemes
- **Start/stop Control**—power cycles compressors based on compressed air demand
- **Load/Unload Control**, also known as constant speed control—motor runs continuously but reduces capacity of the compressor (unloads) when pressure is at satisfactory level
- **Modulating Controls**—varies output to meet flow requirements
- **Dual Control/Auto Dual**—selection of either Start/Stop or Load/Unload. For lubricant-injected rotary screw compressors, auto dual control provides modulation to a pre-set reduced capacity followed by unloading with the addition of an over-run timer to stop the compressor after running unloaded for a pre-set time.
- **Variable Displacement**—alters the percentage of screw compressor rotors working to compress air by redirecting airflow or using

valves. Generally used with inlet valve modulation
- **Variable Speed Drive**—continuously adjusts the drive motor speed to match variable demand requirements

Multiple Compressor Control
- Stop/start, load/unload, modulate, vary displacement, and vary speed schemes are used
- Master control and monitor of all components in the system for improving efficiency
- Network Controls link compressors for signaling with lead compressor sending commands to remaining units
- Pressure/flow and other controllers can add to efficiency improvements
- VFDs can be used in multiple compressor systems, adds control complexity

SAMPLE PROBLEMS

1. Of the following types of air compressors which likely has the highest efficacy (CFM/HP)?
    a.   100 HP screw.
    b.   500 HP two stage reciprocating with intercooling.
    c.   25 HP centrifugal running at 40% load.
    d.   Depends on the temperature of air supplied as input.

2. Which of the following statements regarding air pressures is/are true?
    a.   Power required does not change with pressure of air supplied.
    b.   Power required varies with the square root of the pressure of air supplied.
    c.   Productivity can suffer when air pressure is reduced.
    d.   a and b.
    e.   b and c.

3. A steam turbine has an isentropic expansion enthalpy change of 250 Btu/lb. The actual enthalpy change in going through the turbine is 210 Btu/lb. If the turbine is a backpressure turbine with superheated steam outlet conditions, what is the turbine efficiency?

   a. 48%
   b. 90%
   c. 100%
   d. 84%

4. An electric motor runs a pump with a motor efficiency of 90%.
    Electricity costs $.05/kWh. What is the "breakeven" price of nat-
    ural gas per MCF if a gas motor can do the same job with 30%
    efficiency?
    a. $3.00 per MCF
    b. $4.88 per MCF
    c. $7.95 per MCF
    d. $6.84 per MCF
    e. Not enough information.

5. According to the centrifugal pump laws, pressure in a pumping
    system would vary:
    a. With the square of the RPM of the drive motor.
    b. With the square of the GPM being delivered.
    c. With the cube of the RPM.
    d. With the cube of the GPM.
    e. a and b.

SOLUTIONS TO SAMPLE PROBLEMS

        1. b
        2. e
        3. d
        4. b
        5. e

Chapter 8

Building Envelope

INTRODUCTION

The **building envelope** is defined as the physical separator between the interior and exterior of a building. The building envelope consists of the wall, floors, roofs, doors and any structural openings such as windows. The construction of the building envelope greatly affects its ability to isolate the interior from the exterior.

Topic Areas

Thermal Resistance	Heat Transfer Coefficients
Insulation	Vapor Barriers
Solar Heat Gain	Solar Shading
Thermally Light Facilities	Thermally Heavy Facilities
Conduction Heat Loads	Psychometric Chart
Air Heat Transfer	Water Heat Transfer

THERMAL RESISTANCE

Thermal resistance is the temperature difference across a structure when a unit of heat energy lows through it in unit time. It is the reciprocal of thermal conductance. For materials, thermal resistance is known as the R value and thermal conductance is known as the U value. To increase thermal resistance, one applies insulation. R can be obtained by knowing the conductance, C, for a specified thickness of material.

$$R = 1/C$$

If the conductivity of a material, k, is known, R can be calculated by knowing the conductivity and the thickness of the material in inches.

$$R = t/k$$

INSULATION

There is a wide range of insulating materials. Each property is defined and rated in terms of:
- Cell Structure
- Temperature Use (materials have recommended temp ranges; material breaks down for given temperatures, usually at the high end)
- Thermal Conductivity—the thermal conductivity (K) value should be chosen for mean temperature experienced by the insulation
- Fire Hazard—ratings for flame spread and smoke development
- Forms—blankets, bats, rigid boards, blocks, half-pipe sections, etc.

Common Insulating Materials
- Mineral fiber-rock wood: Mineral fiber insulation made from molten rock. Fairly impervious to heat, used in relatively high temps.
- Fiberglass—most popular, blankets, batts, boards, pipe coverings; temp range is somewhat limited.
- Foam—good K values but not as fire retardant, or not very good K values but good fire retardant; applicable in cold applications.
- Calcium Silicate—good for high temps, made of silica and lime, extremely durable, high thermal resistance.
- Refractories-Ceramic Fiber—used alone or added to fire brick; alumina-silica product.
- Refractories-Fire Brick—made of refractory clay; high temp resistance, low heat storage.
- Others: cellular glass, perlite, diatomaceous earth.

Note—As you add layers of insulation, the resistance values add directly.

Economical Thickness:
As thickness increases, insulation cost increases, loss of heat reduces.
There is an Economical Thickness at *Minimum Cost* to make the solution worthwhile.

SOLAR HEAT GAIN

Solar heat gain refers to the increase in temperature in a space or structure that results from solar radiation. For windows, the solar heat gain coefficient (SHGC) is the fraction of incident solar radiation that passes through a window. It has a value between 0 and 1. A high value signifies high heat gain. The SHGC is influenced by glazing. An uncoated water-white clear glass may have a value above 0.8 while a highly reflective coating on a window might yield a value of 0.2.

THERMALLY LIGHT/HEAVY FACILITIES

The effectiveness of the heating and cooling operations in a building is influenced by the amount of thermal mass in a building along with the presence of equipment that creates internal heat gains. Major equipment, lighting and even people contribute to internal heat gain. If a building is classified as thermally heavy, it is relatively independent of outdoor conditions. Thermally light buildings are greatly influenced by outdoor conditions. The "lighter" the building, the greater the correlation between building energy consumption and Degree-Days.

CONDUCTION HEAT LOADS

Conduction is the transfer of heat through solid objects, such as the ceilings, walls, and floors of a building. Insulation reduces conduction losses. The direction of heat flow is from hot to cold. As temperature difference increases, heat loss increases.
Note that heat transfer also occurs through:
- **Convection**—Heat transferred between a moving liquid or gas and some conducting surface
- **Radiation**—waves (no medium) i.e. heat transfer from the sun

Conduction Heat Transfer Equation

$$q = (A * \Delta T)/\Sigma R$$

Where
    q = heat flow in Btu/hr
    A = area in ft^2
    T = temperature differential in degrees Fahrenheit
    ΣR = sum of the thermal resistance of all layers in ft^2-hr-°F/Btu

Similarly, if using conductance instead of resistance, recall that U = 1/R. Thus,

$$q = U * A * \Delta T$$

Where
　q = heat flow in Btu/hr
　A = area in ft^2
　ΔT = temperature differential in degrees Fahrenheit
　U = thermal conductance in Btu/ft^2-hr-°F

This formula can also be used when looking at Annual Heating Loads for a building. The value of T is replaced with: T = Heating Degree Days * 24 or Cooling Degree Days * 24.

AIR HEAT TRANSFER

Air infiltration can create a large heat flow. The formula for calculating the heat flow caused by air leakage assuming only sensible heating is:

$$q = m * Cp * \Delta T$$

Where
　q = heat flow in Btu/hr
　m = mass flow of air in lbs/hr
　Cp = specific heat of air, 0.24 Btu/lbs-°F
　ΔT = temperature differential, °F

If looking at air leakage for ducts we use,

$$q = 1.08 * cfm * \Delta T$$

Where
　q = heat flow in Btu/hr
　Cfm = duct leakage rate to the outside
　ΔT = temperature differential, °F

Note that the 1.08 comes from:

(cfm * 0.24 * 60 min/hr)/13.5ft^3/lb air = 1.08 * cfm

For a small building, the air leakage rate for the building may be determined from a blower door test. If that value is provided, the heat flow may be determined using:

q = 0.018 * ft^3/hr * T

where ft^3/hr = air leakage rate for the entire building

If we want to know the total heat flow, we must consider enthalpy. For air, we use:

q = CFM * 4.5 * Δh

We can use the psychometric chart to determine the change in enthalpy (Δh).

HEAT TRANSFER COEFFICIENT

The **heat transfer coefficient** (h) for a material is the proportionality constant between the heat flux and the driving force from the flow of heat (temperature difference).

h = q/ΔT

where
    q = amount of heat transferred in Watts/meter^2
    h = heat transfer coefficient in Watts/meter^2-K
    ΔT = difference in temperature

VAPOR BARRIERS

Vapor barriers help prevent water vapor from the interior of a building filtering through the wall and condensing on the warm side of the installation. Polyethylene plastic sheeting is often installed on the inside of a frame wall between the studs and drywall to act as a vapor barrier. This should not be confused with an air barrier whose purpose is to prevent the flow of air. These are often applied in cold climates and can create a "tight" building with little air infiltration. Vapor barriers increase energy efficiency but reduce air infiltration and may create a need for increased make up air.

SOLAR SHADING

Solar shading refers to systems used to control the amount of heat and light from the sun that is admitted to a building. Solar shading devices can reduce the amount of energy required for heating or cooling, optimize the amount of energy used for lighting through daylight harvesting, and enhance indoor comfort.

Solar control and shading can be provided by using:
- Landscape features such as mature trees
- Exterior building elements such as fixed overhangs on south-facing glass
- Glass coatings such as low emissivity glass coatings and window films
- Horizontal reflecting surfaces such as light shelves
- Internal shading such as window blinds

Note that internal shades may positively affect personal comfort but they do not reduce solar gain since the energy has already entered the space.

Window films performance is measured by a **shading coefficient** (SC). If SC = 0, no heat passes and if SC = 1 all heat passes. The fractional heat flow reduction from a window film is given by (1 – SC).

PSYCHOMETRIC CHART

The **psychometric chart** represents the physical and thermal properties of moist air in graphical form. The Psych Chart allows complex problems to be worked out easily, and provides a feel for common HVAC processes of interest.

The standard ASHRAE Psych Chart has a horizontal axis for dry bulb temperature and a vertical axis for humidity ratio in pounds of moisture per pound of dry air. Other parameters on the chart are: relative humidity, wet bulb temperature, enthalpy, specific volume, and saturation temperature. This is discussed further in Chapter 5.

WATER HEAT TRANSFER

The heat transfer equation for water is similar to that for air. The general heat flow equation is:

$$q = m * Cp * T$$

Where
    q = heat flow in Btu/hr
    m = mass flow of air in lbs/hr
    Cp = specific heat of air, 0.24 Btu/lbs-°F
    T = temperature differential, °F

For a flow of water when considering sensible heat only, we can use:

$$q = (GPM * (8.34\ lb/gal)*(60\ min/hr) * (1\ Btu/lb\text{-}°F) * \Delta T$$

In simplified form:

$$q = GPM * 500 * \Delta T$$

ADDITIONAL REFERENCE TERMS

Thermal Equilibrium:
Total heat flow through a system = Heat flow through any part of the system.

Dry Bulb, Wet Bulb and Dew Point temperatures are commonly used to *determine the state of humid moist air*.

Dry Bulb Temperature—T_{db}
The Dry Bulb temperature, usually referred to as air temperature, is the air property that is most common used, basically the ambient air temperature.

It is called *"Dry Bulb"* because the measured air temperature is not affected by moisture content.

The dry-bulb temperature indicates heat content and is shown on the bottom axis of the psychrometric chart.

Wet Bulb Temperature—T_{wb}

The wet bulb temperature is the temperature of air cooled to saturation.

At 100% relative humidity, dry bulb and wet bulb temperatures are equal.

Dew Point Temperature

The dew point is the temperature at which water vapor in the air begins to condense (air becomes 100% saturated). Above this temperature, moisture will stay in the air.

Pickup and Pull-down Times

Pickup and pull-down are the processes of restoring a comfortable room temperature after prolonged inactivity.

If a building has a high thermal quality, it will be better at retaining its original temperature, and will therefore not require a great deal of pickup or pull-down time.

Dew Point Temperature Charts

Dew point temperatures from dry and wet bulb temperatures are indicated in standard charts.

SAMPLE PROBLEMS

1. Thermal conductivity is the reciprocal of the thermal resistance.
    a. True
    b. False

2. What is the approximate density of air at 85°F dry bulb and 50% relative humidity.
    a. 0.0603 lb/ft³
    b. 0.0645 lb/ft³
    c. 0.0687 lb/ft³
    d. 0.0714 lb/ft³
    e. 0.075 lb/ft³

3. What is the cost to make up for the heat loss from a 1000 ft² wall with an R value of 4, in a climate with 3500 Heating Degree Days. The heater is 83% efficient, and fuel costs $5.00 per million Btus.
    a. $1.26/yr
    b. $12.65/yr
    c. $50/yr
    d. $75.50/yr
    e. $126.50/yr

4. Determine the approximate optimum thickness T, in inches, for insulation in a facility where the cost of insulation is $0.10 × T and the cost of the remaining saved energy is $0.40/T.
    a. 0.5 inches
    b. 1.0 inches
    c. 1.5 inches
    d. 2.0 inches
    e. Greater than 4.0 inches

5. A one gpm flow of water can create approximately how many times as great a Btu/hr heat flow as one cfm of air, for the same ΔT temperature difference?
    a. 1.08
    b. 3.412
    c. 108
    d. 341.2
    e. 463

SOLUTIONS TO SAMPLE PROBLEMS

    1. b
    2. d
    3. e
    4. d
    5. e

Chapter 9

CHP Systems and Renewable Energy

INTRODUCTION

Combined Heat and Power (CHP) systems, also known as cogeneration, use engines that combust fuel to generate heat in the process of producing electricity. Renewable energy is generated using the energy from natural sources.

CHP systems and renewable energy offer efficiency improvements and alternatives to traditional sources of energy from the regional electric grid. The details of their operation will be covered in this chapter.

Topic Areas

Topping Cycles	Bottoming Cycles
Combined Cycles	Fuel Selection
Prime Movers	Operating Strategies
Regulations	Codes and Standards
Combined Heat and Power	Distributed Generation
HHV and LHV	Thermal Efficiencies
Solar, Wind, Biomass, and Hydropower	Wind Energy Systems
Solar Thermal and Solar Photovoltaic Systems	

COMBINED HEAT AND POWER SYSTEMS

Combined Heat and Power (CHP)—a system that utilizes a heat generating engine to simultaneously generate electricity and useful heat. CHP is also referred to as a **cogeneration system**.

CHP uses heat that would be wasted in a conventional power plant. Therefore, less fuel needs to be consumed to produce the same amount of useful energy.

Combined Cooling, Heat and Power (CCHP)—system producing electricity, heat and cooling. CCHP is also referred to as a **tri-generation system**.

A CHP system captures the heat energy that is generated when producing electricity. The heat can be used for domestic hot water (DHW) and space heating requirements. A tri-generation system would also utilize the captured heat energy for cooling with the use of an absorption chiller or other heat exchanger equipment.

Conventional central coal- or nuclear-powered power stations convert approximately 33% of their input heat to electricity. The remaining 67% emerges from the turbines as low-grade waste heat with no significant local use and is usually rejected to the environment.

TOPPING CYCLE

Topping Cycle—utilizes the primary energy source to generate electrical or mechanical power. Secondarily, the rejected heat, in the form of useful thermal energy, is supplied.

Equipment Used in Topping Cycle:
- Combustion turbine-generators
- Steam turbine-generator sets
- Reciprocating internal-combustion-engine generators

BOTTOMING CYCLE

A **Bottoming Cycle** utilizes the primary energy source to generate a useful heating process. Secondarily, the rejected heat from the process is then used to generate electrical power.

Example: waste heat from a process directed to waste-heat-recovery boiler that converts this thermal energy to steam, which is supplied to a steam turbine to generate electrical power.

COMBINED CYCLE

Basic Combined (Thermodynamic) Cycle—consists of two power plant cycles.

- **Brayton** cycle—a gas turbine cycle
- **Rankine** cycle—a steam turbine cycle

The gas turbine power plant cycle is the **topping cycle**. Heat and work transfer processes occur in the high temperature region.

The **Rankine steam cycle** takes place at a low temperature and is known as the **bottoming cycle**. Heat is transferred from high temperature exhaust gas to water in a waste heat recovery boiler to create steam.

Combined Cycle Arrangement

A combined cycle utilizes a set of heat engines that convert heat into mechanical energy, to:

1)   Drive electrical generators
2)   Capture waste heat energy

PRIME MOVERS AND FUEL SELECTION

Prime Mover—a machine that converts energy from a fuel into mechanical energy.

Fuel selection is based on CHP application and engine operation.

- **Steam turbine**—converts high pressure steam thermal energy to kinetic energy (energy of the movement of the steam through nozzles) into mechanical power through rotating blades
- **Gas turbine**—thermal energy of pressurized gases create mechanical power through rotating blades
  - Waste gases (residual energy) can be used for heating a facility
  - Mechanical energy can produce electricity with a generator or drive equipment (pumps, compressors, etc.)
  - Fuel is typically natural gas
- **Reciprocating Engines**—provides process heat or steam from engine exhaust
  - Typically for smaller co-generation
  - Heat exchangers may also provide hot water or hot air
  - Fuels used: High speed diesel (HSD), distillate, residual oils, natural gas, liquid propane gas (LPG)
- **Fuel Cells**
  - Molten-carbonate

— Solid oxide fuel cells
- **Other Fuels**
  — Biomass
  — Industrial and Municipal waste
  — Combined gas and solar photovoltaic generation to improve technical and environmental performance

HHV and LHV

Evaluating Fuel Energy Capacity: HHV and LHV
Two methods to evaluate fuel energy capacity:
1) Higher Heating Value (HHV)—the full energy content including all products of combustion
2) Lower Heating Value (LHV)—omits the energy in the water vapor that is formed during combustion of the fuel.
   a. The energy of the water vapor is about 10% of the total energy content.
   b. LHV values are therefore 10% less than the HHV value for the same fuel.

THERMAL EFFICIENCIES

Thermal efficiency in a tri-generation system is defined as:

$$\eta th \equiv \frac{W_{out}}{Q_{in}} \equiv \frac{\text{Electrical Power Output} + \text{Heat Output} + \text{Cooling Output}}{\text{Total Heat Input}}$$

Where
ηth = Thermal efficiency
W_{out} = Total work output by all systems
Q_{in} = Total heat input into the system

Typical tri-generation modeled losses:
- Electricity: 40-45%
- Heat + Cooling: 35-40%
- Heat Losses: 10-15%
- Line Losses: 1-2%

OPERATING STRATEGIES

High Level Strategies
- Energy cost efficiency
- Carbon emission reduction
- Industrial or manufacturing process cost improvement
- Central heat and/or hot water efficiency

Additional Strategies and Applications
- Heat recovery steam generator (HRSG)—uses hot exhaust gases from gas turbines or reciprocating engines to heat water and generate steam for industrial processes.
- Micro CHP—a distributed energy resource (DER), usually less than 5 kWe. Space heat and hot water plus electricity for home or small business.
- Small scale CHP systems—back-up for residential-scale photovoltaic (PV) arrays.

REGULATIONS

United States Department of Energy Goal: CHP will comprise 20% of generation capacity by 2030.

Eight Clean Energy Application Centers established to develop the required technology, lead "clean energy" (combined heat and power, waste heat recovery and district energy) technologies as viable energy options.

DOE/FERC regulate the standards for deployment of combined heat power and distributed generation.

CODES AND STANDARDS

FERC Qualifying Facility (QF)—Cogeneration facility, including any diesel and dualfuel cogeneration facility, is a qualifying facility if it:

(1) **Meets any applicable standards and criteria specified in §§292.205 (a), (b) and (d);** and
(2) Unless exempted by paragraph (d), has filed with the Commission a notice of self-certification pursuant to §292.207(a); or has filed

with the Commission an application for Commission certification, pursuant to §292.207(b)(1), that has been granted.

§292.205 Criteria for qualifying cogeneration facilities.

(a) Operating and efficiency standards for **Topping cycle** facilities

    (1) *Operating standard.* For any toppingcycle cogeneration facility, the useful thermal energy output of the facility **must be no less than 5 percent of the total energy output** during the 12month period beginning with the date the facility first produces electric energy, and any calendar year subsequent to the year in which the facility first produces electric energy.

    (2) *Efficiency standard.*

        (i) For any toppingcycle cogeneration facility for which any of the energy input is natural gas or oil, and the installation of which began on or after March 13, 1980, **the useful power output of the facility plus one half the useful thermal energy output**, during the 12month period beginning with the date the facility first produces electric energy, and any calendar year subsequent to the year in which the facility first produces electric energy, must:

            (A) Subject to paragraph (a)(2)(i)(B) of this section be no less than 42.5% of the total energy input of natural gas and oil to the facility; or

            (B) If the useful thermal energy output is less than 15 percent of the total energy output of the facility, be no less than 45 percent of the total energy input of natural gas and oil to the facility.

        (ii) For any topping cycle cogeneration facility not subject to paragraph (A)(2)(i) of this section, there is no efficiency standard.

(b) Efficiency standards for **bottoming cycle** facilities

    (1) For any bottomingcycle cogeneration facility for which any of

the energy input as supplementary firing is natural gas or oil, and the installation of which began on or after March 13, 1980, the useful power output of the facility during the 12month period beginning with the date the facility first produces electric energy, and any calendar year subsequent to the year in which the facility first produces electric energy **must be no less than 45 percent of the energy input of natural gas and oil for supplementary firing.**

(2) For any bottomingcycle cogeneration facility not covered by paragraph (b) (1) of this section, there is no efficiency standard.

(c) **Waiver.** The Commission may waive any of the requirements of paragraphs (a) and (b) of this section upon a showing that the facility will produce significant energy savings.

(d) *Criteria for new cogeneration facilities.* Notwithstanding paragraphs (a) and (b) of this section, any cogeneration facility that was either not a qualifying cogeneration facility on or before August 8, 2005, or that had not filed a notice of selfcertification or an application for Commission certification as a qualifying cogeneration facility under §292.207 of this chapter prior to February 2, 2006, and which is seeking to sell electric energy pursuant to section 210 of the Public Utility Regulatory Policies Act of 1978, 16 U.S.C. 824a1, must also show:

(1) The **thermal energy output of the cogeneration facility is used in a productive and beneficial manner**; and

(2) The electrical, thermal, chemical and mechanical output of the cogeneration facility is **used fundamentally for industrial, commercial, residential or institutional purposes** and is not intended fundamentality for sale to an electric utility, taking into account technological, efficiency, economic, and variable thermal energy requirements, as well as state laws applicable to sales of electric energy from a qualifying facility to its host facility.

(3) **Fundamental use test.** For the purpose of satisfying paragraph (d)(2) of this section, **the electrical, ther-**

mal, chemical and mechanical output of the cogeneration facility will be considered used fundamentally for industrial, commercial, or institutional purposes, and not intended fundamentally for sale to an electric utility if at least 50 percent of the aggregate of such output, on an annual basis, is used for industrial, commercial, residential or institutional purposes. In addition, applicants for facilities that do not meet this safe harbor standard may present evidence to the Commission that the facilities should nevertheless be certified given state laws applicable to sales of electric energy or unique technological, efficiency, economic, and variable thermal energy requirements.

(4) For purposes of paragraphs (d)(1) and (2) of this section, a new cogeneration facility of 5 MW or smaller will be presumed to satisfy the requirements of those paragraphs.

(5) For purposes of paragraph (d)(1) of this section, where a thermal host existed prior to the development of a new cogeneration facility whose thermal output will supplant the thermal source previously in use by the thermal host, the thermal output of such new cogeneration facility will be presumed to satisfy the requirements of paragraph (d)(1).

DISTRIBUTED GENERATION

Distributed generation has its greatest benefits when scaled to fit buildings or complexes of buildings in which **electricity, heating and cooling are perpetually needed**. Installations include:

• Data centers
• Manufacturing facilities
• Universities
• Hospitals
• Military complexes
• Large multi-family buildings and campuses

Additional Benefits
- Redundancy of power in mission critical applications
- Lower power usage costs
- Ability to sell electrical power back to the local utility

SOLAR THERMAL AND SOLAR PHOTOVOLTAIC SYSTEMS

Solar Thermal—uses thermal energy from the sun to heat water.
Solar Photovoltaic Systems—converts light from the sun to direct current (DC) electricity.

WIND ENERGY SYSTEMS (WIND TURBINES)
(Convert kinetic energy from the wind into electrical power)

Types
- Horizontal Axis
- Vertical Axis

Components
- Rotor—includes blades to convert wind energy to rotational energy at a low speed
- Controls—converts low-speed rotational energy to high speed rotational energy for electricity generation
- Generator—creates electricity form rotational energy
- Structure—Tower and related mounting apparatus

BIOMASS

Biomass
Plant or animal waste that can be refined to natural biofuels.
- Wood
- Sugarcane
- Corn
- Agricultural waste
- Municipal waste

Biofuel Engine CHP Facilities
- Use an adapted reciprocating gas engine or diesel engine depending upon specific biofuel.
- Very similar in design to a Gas engine CHP plant.

Key Benefit
Reduced hydrocarbon fuel consumption and carbon emissions.

Wood Gasifier CHP
Wood pellet or wood chip biofuel creates a gas used to power a gas engine.

HYDROPOWER

Generation of electricity from the kinetic energy (movement) of water.
- Components
  — Dam—opened or closed to control water flow.
  — Turbine—blades rotate from water motion.
  — Electric Generator—creates electricity form turbine rotational energy.

- Advantages
  — Inexpensive after dam is built (water flows naturally).
  — Water is a clean, renewable fuel source from snow and rainfall.
  — Reservoirs may offer recreational options.

- Disadvantages
  — Damming rivers may destroy or disrupt wildlife.
  — Can cause low dissolved water oxygen, potentially harmful to wildlife.

SAMPLE QUESTIONS

1.   In Topping Cycle cogeneration systems, Electricity or mechanical power is produced first; then heat is recovered to meet the thermal loads of the facility.

    a.   True

    b.   False

    c.   Not enough information

2.   The FERC requirement for QF status, assuming thermal output is 15% or greater of the total energy output, is:

    a.   (power output +1/2 useful thermal output)/energy input > 42.5%

    b.   (power output + 1/2 useful thermal output)/energy input > or = 42.5%

    c.   (power output + 1/2 useful thermal output)/energy input < or = 42.5%

    d.   (power output + 1/2 useful thermal output)/energy input = 42.5%

3.   The ratio of the LHV to HHV for natural gas is approximately:

    a.   0.8

    b.   0.9

    c.   1.0

    d.   0.7

4.   The CEM Hotel has been approached by a cogeneration developer who claims that they can reduce the hotel's energy costs. They plan to install a 100 kW packaged cogeneration system with the following specifications listed below.

Specifications:

Output: 100 kW

Electrical Efficiency: 25%

Total Efficiency: 75%

Fuel Cost: $0.40/therm

Current electric Cost: $0.06/kWh

Existing boiler efficiency: 70%

Hotel Hot Water Usage: 20,000 gallons/day (70 F Rise)

What is the hourly fuel consumption in million Btu/h and cost?

    a.   1.365 million Btu/h, $5.46/hr

    b.   2.312 million Btu/h, $3.41/hr

    c.   0.365 million Btu/h, $6.58/hr

    d.   4.219 million Btu/h, $3.46/hr

5.   A back pressure turbine (assume 100% efficient turbine) receives
      10,000 pounds per hour of steam at 100 psig and 112°F superheat.
      Outlet conditions of the steam are 15 psig, saturated. If the electri-
      cal generator is 95% efficient, how much electrical power is gener-
      ated?
      a.   176 kW
      b.   195 kW
      c.   211 kW
      d.   244 kW
      e.   325 kW

SOLUTIONS TO SAMPLE QUESTIONS

      1.   a
      2.   b
      3.   b
      4.   a
      5.   d

Chapter 10

Fuel Supply and Pricing

INTRODUCTION

The procurement of energy includes purchasing electricity, oil, gas, and other fuels for use at a facility. Energy procurement in both regulated and deregulated (competitive) marketplaces differs based on state government regulations and utility tariffs that determine cost structures. Therefore, an understanding of procurement principles and billing line items is a necessity for energy managers to manage commodity purchases, as well as determine return on investments for energy efficiency projects. Levels of risk in speculative commodity futures purchasing should be considered as well.

Topic Areas
Procurement of Natural Gas
Supply and Demand Impact on Pricing
Fuel Price Risks
Electricity as a Commodity
Procurement of Oil
Evaluating Supply Options
Trends in Deregulation around the World
Selection of Energy Supplier in a Deregulated Market

GENERAL

Energy Procurement Factors:
- Fuel Availability
- Purchase price
- Contract terms, clauses
- Volume commitments
- RISK tolerance
- Environmental (emissions, etc.)

PROCUREMENT OF NATURAL GAS

Two main natural gas price components:

1) Cost of Natural Gas itself ("raw material," known as **Supply cost**)
2) **Basis Cost** to transport the gas from source to LDC; Basis Cost is the expected difference between the price at the delivery point at the natural gas hub and the local utility (LDC).
   - Pipeline tolls
   - Storage charges   — System losses
   - Capacity requirements   — Supplier margin

Natural Gas Procurement Strategies in De-regulated Markets
- Unbundle and separately purchase commodity, supplier fees, and basis fees
- Review pricing history for supply and basis costs, near and long term trends, and seasonality
- Implement hedging strategies to mitigate risk (see Fuel Price Risks below)

SUPPLY AND DEMAND IMPACT ON PRICING

- Reduced natural gas inventories increase supply gas supply costs
- Electricity market pricing tacks closely to natural gas price trends

FUEL PRICE RISKS

Fuel price risk: uncertainty of future price fluctuations of the energy commodity. Potential factors that influence energy supply costs:
- Global demand
- National demand
- Severe weather events that threaten production or transport
- Variations in actual temperatures vs. forecasted

Risk mitigation strategies: reduce or eliminate exposure to fluctuating fuel costs by implementing hedging strategies:
- Stagger purchases over time (dollar cost averaging)
- Purchase both fixed rate and variable (floating) contracts based on seasonality and business requirements

ELECTRICITY AS A COMMODITY

Electricity Pricing Components
- Electricity Generation: electricity supply generated by power plants
- Electricity Transmission: use of high voltage power lines to carry power to local utilizes (LDCs) for distribution
- Electricity Distribution: the "last mile" of power for delivery to customers

Electricity Generation and **Transmission** are the pricing components available for contract purchasing in U.S. deregulated states.

- Basic Generation Services **($/kW)**; also known as BGS **Capacity**
  — **Generation** ($/kW)
  — **Transmission** ($/kW)
- Basic Generation Services **($/kWh)**; also known as BGS **Energy**

Electricity Distribution
- Local Distribution Company (LDC) Delivery
  — Customer service charge ($)
  — Demand charge ($/kW)
  — Energy Delivery charge ($/kWh)
  — Other fees ($/kWh)
    - Societal benefits charge (energy efficiency fund)
    - Reconciliation charge (adjustments)

PROCUREMENT OF OIL

Oil Price Components
- Oil Drilling costs
- Storage
- Transportation
- Delivery

EVALUATING SUPPLY OPTIONS

Comparison of fuel choices to satisfy the needs of a facility or an industrial process:

- Obtain commodity price quotes (supply and delivery costs)
- Convert $/unit to $/kBtu equivalent for each supply option
- Compare supply options

Contract Terms
- Term contract length
- Cancellation policy and penalties
- Commodity Usage
  — Minimum energy use or volume commitments
  — Natural gas heat: Firm (continuous use) or Interruptible (part-time use of other commodity for heat)
- Responsibilities of parties

TRENDS IN DEREGULATION AROUND THE WORLD

- Shift in utility responsibilities
- Growth of distributed generation
  — Solar photovoltaics
  — On-site co-generation
  — Microgrids
- Distributed resource management, including demand response
- Smart grid technology
- Energy management driven by "big data"
- Cybersecurity concerns and solutions
- Energy storage
  — Fuel cells
  — Off-hour ice generation for chilled water use
- Clean power initiatives and governance
- Blurring Regulatory

SELECTION OF ENERGY SUPPLIER IN A DEREGULATED MARKET

Electricity and Natural Gas Commodities
Regulated Energy Supply
    Local utility companies provide every aspect of electric and gas services

Deregulated Energy Supply
   Customers choose their **energy supplier**
- Energy supply offers a competitive market
- Generation suppliers licensed by a Public Utility Commission or equivalent
- Local utility (LDC) remains responsible for service delivery
- Commercial, Industrial, and Residential
- State utility commission oversight of suppliers

Suppliers and Agents
- Energy Suppliers—contracting entity
- Energy Aggregators—multiple customers part of large contract
- Energy Brokers—agents for multiple suppliers
- Energy auction platform—bidding process for multiple energy suppliers

Energy Contract Factors and Potential Benefits
- Cost reduction
- Competitive pricing from multiple suppliers
- Risk Mitigation
  - Budget certainty vs. additional savings
  - Variable and fixed rate plans
  - Multiple contract term lengths
  - Review sort and long term pricing trends

SAMPLE PROBLEMS

1.   Number 2 fuel oil at $0.95/gal contains 140,000 Btu/gal. If this oil can be burned with an efficiency of 75%, what is the point of use cost of the oil heat?
     a. $5.05/MMBtu
     b. $9.05/MMBtu
     c. $10.05/MMBtu
     d. $14.05/MMBtu
     e.   No answer above is close

2.   Which of the following fuels has the highest percentage of hydrogen and the lowest percentage of carbon?
     a.   Natural gas

    b. Propane

    c. Wood

    d. Pentane

    e. Coal

3.   After the "gathering" operation, natural gas is processed before being put into an interstate pipeline. Which of the following is not a part of that processing?

    a. Removing moisture

    b. Enriching by injecting small amounts of fuel oil

    c. Compressing to a higher pressure

    d. Removing hydrogen sulfide

    e. None of the above—they are all part of the processing.

4.   Landfill gas typically has:

    a. More Btu content than natural gas

    b. Less Btu content than natural gas

    c. About the same Btu content as natural gas

    d. Almost no Btu content

    e. None of the above

5.   A "marketer" of electricity:

    a. Owns the electric energy being sold

    b. Operates as a consultant and does not own the electric energy

    c. Works for the state PUC, PSC or similar agency

    d. Must clear each purchase or sale with FERC

    e. a and d

SOLUTIONS TO SAMPLE PROBLEMS

    1. b

    2. a

    3. b

    4. b

    5. a

Chapter 11

Building Automation and Control Systems

Technology advancements have provided energy managers with greater visibility of energy use, demand loads, and operational characteristics with the use of automation and control systems. Hardware to access energy monitoring points can be installed throughout a facility and accessed by using on-site energy management software, or via software as a service (SaaS) from a provider via the Internet.

This chapter covers key topic areas to familiarize energy managers with core automation and control concepts for commercial, industrial, and large multi-family buildings.

Topic Areas

Energy Management Strategies	Terminology
Basic Controls	PID Controls
BACnet & LON	Signal Carriers
Power Line Carriers	Direct Digital Control
Distributed Control	Central Control
Optimization Controls	Reset Controls
Building Control Strategies	Communication Protocols
Expert Systems	Artificial Intelligence
Self-Tuning Control Loops	Energy Information Systems
TCP/IP	Internet, Intranets and WWW
BAS Systems	Web Based Systems

ENERGY MANAGEMENT STRATEGIES

Energy managers employ strategies to address energy efficiency of facilities, ideally in line with corporate objectives such as:
*   Cost containment
*   Occupancy comfort

- Greenhouse gas reductions
- Fuel conservation
- Water conservation
- Safety

Facility-wide energy management plans are developed to employ strategies to:
- Implement near-term lower cost/complexity energy reduction solutions while planning for longer term projects
- Collect, analyze, and take action using real-time energy data and benchmark reference data
- Compliment the use of in-house staff with industry expertise as needed

Building Automation Systems (BAS)

Building Automation Systems (BAS) are centralized networks of hardware and software to monitor and control equipment in commercial, industrial, and institutional facilities.
- Enables comfort and safety of building occupants
- Ensures the operational performance of the facility
- Installed in new buildings or as a retrofit (addition or replacement of older system) in existing facilities

BAS Functionality
- Equipment control and monitoring
- Operational scheduling
- Data collection and analysis
- Integration with other control and security systems
- Alarming
- Remote accessibility

Mechanical, electrical, and plumbing (MEP), HVAC, and other building systems controlled by a BAS:
- Chillers
- Boilers
- Air Handling Units (AHUs)
- Roof-Top Units (RTUs)
- Fan Coil Units (FCUs)
- Heat Pump Units (HPUs)

- Variable Air Volume (VAVs) units
- Lighting

Additional functionality addressed by a BAS:
- Demand load monitoring
- Security, including close circuit video (CCTV)
- Card and keypad access
- Fire alarm system
- Elevators/escalators status
- Plumbing and water monitoring

TERMINOLOGY
- A **Building Automation System** (BAS) is also referred to by other names, often used interchangeably:
  — Building Automation and Control Systems (BACS)
  — Building Control System (BCS)
  — Building Management System (BMS)
  — Energy Management Control Systems (EMCS)
- **Energy Management System** (EMS): a BAS with emphasis on energy metering and/or monitoring to alert, report, and/or adjust operation
- **Controls**: discrete devices that control particular pieces of equipment or systems
- **Direct Digital Control** (DDC): the communication method used in modern devices (hardware and software) to control building systems; DDCs form the automation system
- **Programmable Logic Controller** (PLC): the controller programming language per IEC 61131-3 standard
- **Smart (Intelligent) Building**: a building equipped with a data-rich BAS

BASIC CONTROLS

The **simplest control** is **on/off**, in which user control or a self-actuating mechanism invokes a desired change in operation.
- **Manual Systems**—on/off switches, variable settings (light dimmer, e.g.)

- **Basic Controls**—timers, e.g.
  - — Account for daylighting using sensors
  - — Timers for night setbacks

All controllers have a set of common characteristics:
- Input signal
- A device to implement the control algorithm
- A process that is controlled
- An output that is used to control the process

Most controls are closed loop in which there is an **error signal** that is developed from the difference between the input and the feedback of the output.

- **Programmable controllers with sensors for processes**
  - — Buildings with 100+ control points; 100,000 ft² or more
  - — Manage demand to not exceed a set peak
  - — Monitor CO_2 level in exhaust to adjust air intake, e.g.
- **Energy Management Control Systems** (EMCS)
  - — Buildings with several to 1,000s of control points
    - Light dimming
    - Duty cycling
    - Combustion control
    - Surge protection/turn-off loads on outage to avoid power surge when service is restored
    - Temperature reset based on heating/cooling demand
    - Fire safety
    - Equipment Maintenance
    - Report generation
  - — Issues to address for EMCS success:
    - Define energy management control requirements
    - Determine EMCS equipment and software
    - Assess vendors carefully before purchasing decision
    - Develop plan to optimize energy efficiency, unless part of the EMCS

DIRECT DIGITAL CONTROL (DDC)

DDC is the method by which a programmed controller receives information (inputs) and then communicates to a device to alter the device's operation.

Most systems now use direct digital control.

- Electronic sensors feed information to the BAS, which in turn sends control signals to the appropriate valves, dampers, and other devices.
- Faster, more flexible, and more accurate than older pneumatic controls

DISTRIBUTED CONTROL

Distributed control is the method by which multiple controllers communicate to multiple devices, either locally or remotely over the selected signal carrier.

- Remote devices are connected to a field interface device (FID) that interprets commands
- Fully distributed systems use FIDs that receive sensed information and perform control functions independently

CENTRAL CONTROL

Central Control allows for master control across an entire facility by collecting data from all subsystems.

OPTIMIZATION CONTROLS

Optimization Controls are intended to improve the performance of a particular process or equipment element, possibly as a retrofit to a pre-existing BAS.

RESET CONTROLS

Reset Controls are used to change setpoint values for device operation based on variations of conditions, such as a drop in ambient temperature as an example.

Proportional control: the manipulated variable is directly proportional to the magnitude of the error signal. This mode of control is common as it drives the process in a simple linear way.

Integral control: the value of the manipulated variable is changed at a rate that is proportional to the error signal: the error is averaged over time.

Derivative controller: the manipulated variable is proportional to the rate of change of the error signal. Faster system response to inputs is achieved with this control mode; usually used in conjunction with proportional control.

PROPORTIONAL, INTEGRAL, DERIVATE (PID) CONTROLS

A **PID Controller** is a control loop feedback mechanism commonly used in industrial control systems. A calculation is performed between the measured output value and a reference value (setpoint), and adjustments are made to produce the desired output value.

- *Proportional, integral,* and *derivative* algorithms are used to calculate a reference value as the weighted sum of each algorithm
- Proportional, integral, and derivative setpoints are initially set, subsequently adjusted ("tuned") for an application, and periodically re-tuned to maintain the desired operation
- Incorrect PID settings can cause a system to continuously "chase" the desired output. Outputs that change quickly to reach and exceed setpoints will cause an increase in energy consumption and potential mechanical equipment failures, equivalent to "short-cycling" operation

BUILDING CONTROL STRATEGIES

Building Control Strategies and use of a BAS are based on saving energy while providing increased user comfort at reduced operational cost.

Strategies Include
- Centralized monitoring and control for all building service systems
- Include building life cycle in BAS design and implementation
- Realize benefits of information exchanged between existing building systems
- Combine independent systems to enhance and maximize life-cycle cost and functionality benefits

- Integrate security systems, including fire alarm system
- Migrate to use a common, open communication protocol for operational simplicity and reduced maintenance costs
- Leverage existing IT and communication systems for BAS operations
- Design for comprehensive automation to achieve intelligent building operation
- Support tenant/occupant requirements for productive working environments, by considering variations in building uses
  — Sunlit facing building sections vs. little or no sunlight impact
  — Personnel activities and expected attire
- Incorporate potable and waste water management in BAS planning
- Account for emergency alarm conditions to be detected and communicated

COMMUNICATIONS PROTOCOLS

Communications protocols are a set of messages to be sent and received between computers and devices to control equipment, obtain data, and report status. Protocols are fundamentally computer languages that allow devices to operate as a system.
- BACNet is a protocol standard (see below).
- Digital Addressable Lighting Interface (DALI) is a control method for building lighting. DALI devices interface with light ballasts through a dedicated cable that is shared among all devices.
- X10 is a language and protocol that uses a power line carrier (see below) on existing wiring.

BACNET PROTOCOL

The Building Automation and Control Networks (**BACnet**) protocol was developed by **ASHRAE** (Standard 135-2016) to meet building automation and control system communication needs. Communication services and protocols are defined for equipment that controls building systems.

Used in:
- HVAC/R Systems

- Lighting Control
- Energy Management Systems
- Physical Access Control
- Fire and Other Safety/Security Systems
- Elevator Monitoring Systems

BACnet communications are between:
- Direct digital controllers
- Application-specific controllers
- Individual equipment controllers
- Head-end computers
- Mobile and cloud-hosted devices

LON

LONWorks (LON: Local Operating Network), similar to BACnet operation, is a network protocol that uses specific microprocessor and communication devices for building automation and control applications.
- LONWorks defines messaging content and structure to be communicated between devices
- LONTalk addresses the communication aspects for commands and information exchanged between devices.
- Control networking standard CEA 709.1-D-2014 (ANSI)
- Built by Echelon Corporation

SIGNAL CARRIERS

Signal Carriers are the paths by which communication protocol messages travel, including twisted pair copper wires, fiber optic cable, and RF (radio frequency/wireless), and power lines.

Power Line Carriers utilize existing power lines to send and receive data between BAS devices. The same power lines simultaneously carry Alternating Current (AC) electric power transmission or distribution.

EXPERT SYSTEMS AND ARTIFICIAL INTELLIGENCE

Expert systems, or knowledge-based systems, are computerized systems and/or devices that have incorporated specific subject matter knowledge into a decision-making process.

Expert systems incorporate the principles of **Artificial Intelligence** (AI) by which a machine continually calculates best actions to take based on changing conditions (inputs), beyond a simpler comparison of discrete values.

- Energy Management Systems can utilize expert system programming to manage building energy assets
- An interactive text or graphical user interface (GUI) asks detailed questions to gather building and use information
- Expert systems programming used to help solve specific operational problems

SELF-TUNING CONTROL LOOPS

Self-Tuning Control Loops are feedback control systems in which the user can issue a command to recalculate setpoint parameters, or "tune" the system for a given process with the control function disabled.

- Controller tuning function exercises the process until satisfactory numbers of trial input and output data have determined the behavior of the process
- Tuning includes updating the controller's P, I, and D tuning parameters
- An Auto-Tuning Controller is similar to a self-tuning control loop, except it performs a tuning operation once and then performs control functions using the computed parameters
- Many PID controllers include both auto-tuning and self-tuning options

ENERGY INFORMATION SYSTEMS

Energy Information Systems provide centralized databases of building standards, safety codes, specific equipment operating specifications, and related information that can be leveraged for day-to-day energy management. Internet access to energy information, whether free or by subscription, offers low-cost access to updated data.

INTERNET, INTRANETS, WWW, AND TCP/IP

Internet access and communications provide the means to collect system wide data from multiple facilities and use updated energy information to best manage energy performance.

Intranet communications are the local area network (LAN) connectivity methods to control equipment and exchange data. Protocols used:

- Ethernet, used in the Transmission Control Protocol and Internet Protocol, commonly known as **TCP/IP**
- Wireless (IEEE 802.11g, commonly known as Wi-Fi)
- Modbus, uses RS-232 or RS-485 serial hardware interfaces

WEB BASED SYSTEMS

Building automation system capabilities are available via the Internet, providing economies of scale to manage multiple facilities with a common platform

- **Software as a Service (SaaS)** business model, working in conjunction with hardware vendors to access on-site control and data collection devices
- Centralized management and control systems
- Seamless software upgrades and new features
- Connectivity to utility operated demand response systems

SAMPLE QUESTIONS

1. In a VAV system, when using mechanical cooling, it is best to deliver supply air at the maximum temperature possible and, when using outdoor air for cooling, deliver air at the lowest possible temperature.
    a. True
    b. False

2. When the Heating Hot Water supply temperature is reset as the facility heating needs change, we refer to this strategy as:
    a. Heating Hot Water Outdoor Air Reset
    b. Night Setback
    c. Boiler Optimization
    d. Reheat coil reset
    e. None of the above

3.   In a DDC control system, distributed processing means:
     a.   the algorithms for central control are located with the host computer
     b.   the network carries the central station commands to the individual control panels for execution
     c.   the decision-making and control process is largely handled by the local control panel
     d.   All of the above

4.   A system that has dirty condenser coils will have little effect on the facility energy consumption providing that the following energy-savings strategies are already in place:
     a.   duty-cycling of supply fans
     b.   demand shedding of non-essential loads
     c.   chilled water reset
     d.   optimum start-stop programs
     e.   none of the above

5.   It is possible to schedule lighting loads (turn on and off at specific times) for energy reduction by using the following:
     a.   7 day timers
     b.   photocells
     c.   occupancy sensors
     d.   DDC control systems
     e.   a and d

SOLUTIONS TO SAMPLE QUESTIONS

     1.   a
     2.   a
     3.   c
     4.   e
     5.   e

Chapter 12

High Performance Buildings

INTRODUCTION

Green Building is a concept in which the environment is considered when planning the complete life cycle of a building. Rating systems, such as LEED, ENERGY STAR, and Green Globes® guide builders to achieve an energy efficient, cost-effective, and environmentally friendly design.

Topic Areas

Green Buildings	Energy and Atmosphere
USGBC	Materials and Resources
Sustainable Design	Indoor Environmental Quality
LEED CI	Portfolio Manager
LEED Certification	ENERGY STAR Rating
LEED O&M	ENERGY STAR Label
LEED NC	Green Globes
LEED CS	ASHRAE 90.1 Energy Cost
Certified, Silver, Gold,	Budget Method
and Platinum	ASHRAE Standard 189
Water Efficiency	ASHRAE Green Guide

GREEN BUILDINGS

Environmental Considerations Drive:
*   Planning
*   Building design
*   Construction
    — Building Materials
    — Impact on the property and surroundings
*   Operations
    — Energy use
    — Water use
    — Indoor air quality

SUSTAINABLE DESIGN

Principles and Objectives
- Lower the adverse impact on the environment
- Serve the health and comfort of building occupants
- Reduce fossil-fuel consumption
- Improve building operational efficiencies
- Consider the entire building life-cycle

U.S. GREEN BUILDING COUNCIL (USGBC)

The U.S. Green Building Council (USGBC, a 501(c)(3) nonprofit organization) was founded in 1993 to develop a rating system for green building design, construction, and efficient operation. The result of those efforts is the **Leadership in Energy and Environmental Design (LEED)** practice in widespread use today.

LEED is a third-party verification system and design tool for sustainable structures around the world.
- LEED acts as a framework for project team decision-making when planning the complete lifecycle of a building
- All Building Types—Commercial, Industrial, Residential
- Project Team Management has discretion to determine and choose the LEED Certification system to use
- Buildings are certified using a points system for sustainability topic areas under each LEED Ratings System

LEED is NOT:
- A performance measurement tool
- Climate-specific
- Energy-specific

LEED v4 Rating Systems Overview
- Launched November 2013
- Includes new project types and associated requirements; more specialized, flexible vs. previous version LEED 2009 (LEED v3)
- LEED v4 areas of development and improvement
  — Material composition and their effect on human health and the environment

— Performance-based approach to indoor environmental quality for occupant comfort
— Smart grid benefits; credit for demand response program participation
— Water efficiency evaluation based on total building consumption

• Ten LEED 2009 rating systems were incorporated into *five LEED v4 rating systems*:
  — LEED for Building Design and Construction **(BD+C)**
  — LEED for Interior Design and Construction **(ID+C)**
  — LEED for Building Operations and Maintenance **(O+M)**
  — LEED for Neighborhood Development **(ND)**
  — LEED for Homes Design and Construction

LEED v4 Rating Systems
• *LEED BD+C:* Building Design and Construction
  — BD+C: New Construction
  — BD+C: Core and Shell
  — BD+C: Data Centers
  — BD+C: Healthcare
  — BD+C: Hospitality
  — BD+C: Retail
  — BD+C: Schools
  — BD+C: Warehouses and Distribution Centers

• *LEED ID+C:* Interior Design and Construction
  — ID+C: Commercial Interiors
  — ID+C: Hospitality
  — ID+C: Retail

• *LEED O&M:* LEED for Operations and Maintenance
  — O+M: Existing Buildings
  — O+M: Data Centers
  — O+M: Hospitality
  — O+M: Retail
  — O+M: Schools
  — O+M: Warehouses and Distribution Centers

- *LEED ND*: Neighborhood Development
  — ND: Plan
  — ND: Built Project

- *LEED* for Homes Design and Construction
  — Homes and Multifamily Low-rise: single family homes and multifamily buildings between one and three stories.
  — Multifamily Midrise: multifamily buildings between four and eight stories.

LEED 2009 Rating Systems
- Open registration closed October 31, 2016
- Maintenance on-going for active LEED 2009 projects
  — LEED NC—New Construction
  — LEED NC-R—New Construction Retail
  — LEED CS—Core & Shell
  — LEED HC—Healthcare
  — LEED Homes (2008)
  — LEED Schools
  — LEED Existing Buildings
  — LEED CI—Commercial Interiors
  — LEED CI-R—Commercial Interiors Retail
  — LEED ND—Neighborhood Development

LEED v4 Minimum Program Requirements (MPRs)
- Required for LEED certification eligibility
- Defines types of buildings that LEED was designed to evaluate
- Provide guidance, reduce complications, protect the integrity of the LEED program

  — Must be in a *permanent location on existing land* for evaluation in the context of the surroundings; use of artificial land masses for buildings are not eligible.

  — Must use reasonable LEED boundaries to *include all contiguous land associated with the project* to support typical operations.

  — Must comply with *project size requirements*, such as minimum square footage and function, for different LEED rating systems.

LEED v4 Certification: Point Credit System

Performance credit system allocates points based on potential environmental impacts and human benefits.

- US EPA environmental impact categories for the Tools for the Reduction and Assessment of Chemical and Other Environmental Impacts (TRACI)
- Environmental-impact weighting scheme developed by the National Institute of Standards and Technology (NIST).

LEED v4 offers *110 possible base points* distributed across credit categories:

- **LT**: Location and Transportation
- **SS**: Sustainable Sites
- **WE**: Water Efficiency
- **EA**: Energy and Atmosphere
- **MR**: Materials and Resources
- **EQ**: Indoor Environmental Quality
- **IN**: Innovation
- **RP**: Regional Priority

- **Additional Credits** are offered in different rating systems. Examples:
  — Integrated Process
  — Reduced Parking Footprint
  — Open space
  — Light pollution reduction
  — Alternative transportation—bicycle storage and changing
  — Water metering
  — Enhanced Refrigerant Management

- **WE**: Water Efficiency
  — Outdoor water use reduction
  — Indoor water use reduction
  — Building-level water metering
  — Outdoor water use reduction
  — Indoor water use reduction
  — Cooling tower water use
  — Water metering

- **EA**: Energy and Atmosphere Credits—categories for multi-family (MF) and single family (SF) homes
  — Minimum energy performance (MF and SF)
  — Energy metering (MF and SF)
  — Education of homeowner, tenant, or building manager (MF and SF)
  — Efficient hot water distribution system (MF and SF)
  — Advanced utility tracking (MF and SF)
  — Home size (SF)
  — Annual energy use (SF)
  — Active solar-ready design (SF)
  — HVAC Start-up credentialing (SF)
  — Building orientation for passive solar (SF)
  — Air infiltration (SF)
  — Envelope insulation (SF)
  — Windows (SF)
  — Space heating and cooling equipment (SF)
  — Heating and cooling distribution systems (SF)
  — Efficient domestic hot water equipment (SF)
  — Lighting (SF)
  — High-efficiency appliances (SF)
  — Renewable energy (SF)

- **MR**: Materials and Resources Credits
  — Storage and collection of recyclables
  — Construction and demolition waste management planning
  — Building life-cycle impact reduction
  — Building product disclosure and optimization—environmental product declarations
  — Building product disclosure and optimization—sourcing of raw materials
  — Building product disclosure and optimization—material ingredients
  — Construction and demolition waste management

- **EQ**: Indoor Environmental Quality
  — Minimum indoor air quality performance
  — Environmental tobacco smoke control
  — Green cleaning policy

— Indoor air quality management program
— Enhanced indoor air quality strategies
— Thermal comfort
— Interior lighting
— Daylight and quality views
— Green cleaning—custodial effectiveness assessment
— Green cleaning—products and materials
— Green cleaning—equipment
— Integrated pest management
— Occupant comfort survey

Certification Levels
• **Certified**: 40–49 points
• **Silver**: 50–59 points
• **Gold**: 60–79 points
• **Platinum**: 80 points and above

Credit Weighting Process—The weighting process has three steps:
• A collection of reference buildings are used to estimate the environmental impacts of any building seeking LEED certification in a designated rating scheme
• NIST weightings are used to judge the relative importance of these impacts in each category
• Data regarding actual impacts on environmental and human health are used to assign points to individual categories and measures

Green Building Information Gateway (GBIG)
• Connects green building efforts and projects from all over the world
• Provides searchable access to a green building-related database from multiple sources
• Offers information about LEED projects

WATER CONSERVATION AND EFFICIENCY OVERVIEW

Water conservation includes the effective delivery of water to end users, using the sufficient amount of water to perform daily tasks for residences and businesses, and maintaining equipment and systems to prevent waste.

Water Efficiency and Conservation Measures

- Water Metering: accurate use data and direct billing compels users to conserve and suppliers to maintain delivery systems
  — Source metering (Utility)
  — Public-use (free) water metering (Utility)
  — Service-connection metering (Users)
  — Fixed-interval meter reading (Utility)—track against source and service connection data

- Water Accounting and Loss Control: account for water use from source to end use to detect and address system losses
  — Water accounting system
  — Analyze non-account/metered water
  — System audit
  — Leak detection and repair strategy
  — Employ automated sensors/telemetry
  — Loss-prevention program—on-going system maintenance

- Costing and Pricing: Utility establishes value and bills users accordingly
  — User charges, metered rates
  — Perform cost analysis by use, seasonality
  — Define service classes
  — Create incentives for conservation

- Water Conservation Education of End User Customers

- Water-Use Audits: Use-specific
  — Large-landscape audits
  — Selective end-use audits.

- Retrofits—upgrade equipment to conserve water flow
  — Toilets
  — Showerheads
  — Bathroom and kitchen faucets

- Pressure Management in the distribution system

- Landscape Efficiency—large projects

- Replacements and Promotions
  — Rebates and incentives promotions—residential and other use
  — Promotion of new technologies

- Water Reuse and Recycling

ENERGY STAR

ENERGY STAR is a U.S. Environmental Protection Agency (EPA) voluntary program to help building stakeholders implement energy efficiency measures to reduce energy costs and protect the environment.

ENERGY STAR labeled buildings use an average of 35% less energy and generate 35% fewer greenhouse gas emissions than similar type buildings

To maintain consumer trust and improve the oversight of ENERGY STAR certified products, homes, and commercial facilities, EPA has implemented third–party certification requirements and testing.

- Established by the EPA in 1992 under authority of the Clean Air Act Section 103(g) that includes non–regulatory strategies and technologies for reducing air pollution.
- Promotes use of energy–efficient products and buildings that includes labeling to reflect certain criteria have been met.
- Buildings earn ENERGY STAR certification in a similar manner to products
- Twenty-one (21) types of buildings are evaluated under ENERGY STAR

ENERGY STAR PORTFOLIO MANAGER

Portfolio Manager is a software program for determining a building's overall energy efficiency.

- A scoring system ranks buildings by percentile relative to similar buildings nationwide (1-100)
- Evaluates energy and water use, and greenhouse gas (GHG) emissions

EnergyStar Evaluation Parameters
- Operating conditions
- Regional weather data
- Building size
- Building use
- Number of occupants

EnergyStar Portfolio Manager Score Calculation
- An algorithm estimates source energy use across a range of efficien-

cy levels
- Compares entered energy data to estimated efficiencies to determine a building rank relative to comparable buildings
- Accounts for weather variations and operational differences between buildings
- Independent verification is required to rank building with a score of 75 or greater

ENERGY STAR RATING AND LABEL

ENERGY STAR Building Certification
- ENERGY STAR score of 75 or higher is required to earn certification (top 25 percentile)
- Reflects building efficiency and cost reduction while retaining performance
- Lower greenhouse gas (GHG) emissions
- Certification is valid for 1 year
- Verification by a professional engineer or registered architect is required

GREEN GLOBES®

Green Globes is a nationally recognized green rating assessment, guidance and certification program for sustainable building design, construction and operation, developed and offered by the **Green Building Initiative® (GBI)** (501(c)(3) nonprofit organization).

Serves the building stakeholder and sustainability communities and is applied to:
- New Construction
- Existing Buildings
- Interiors

Objectives
- Best practices guidance for green construction and operations
- Identifies opportunities and provides effective on-line tools
- Realize sustainability goals for new construction projects, existing buildings and interiors

- Provides in-depth support for improvements ideally suited to each project
- Building owner and facility manager building knowledge is leveraged with Green Globes real-time customer support
- Third-party assessors available throughout the certification process
- Comprehensive assessment roadmap

Program Benefits
- Reduce operating costs
- Qualify for tax incentives
- Meet government regulations
- Attract and retain employees
- Increase a property's marketability

ASHRAE 90.1 ENERGY COST BUDGET METHOD

A building and its systems must satisfy ASHRAE Building Standard 90.1 compulsory conditions for compliance before it can be considered for LEED certification.

Project teams must demonstrate compliance the most rigorous criteria possible:
- All of the prescriptive provisions in Standard 90.1
- The requirements in the local energy code
- The Energy Cost Budget Method defined in Section 11 of Standard 90.1

Energy Cost Budget (ECB) Method is an alternative approach to demonstrating compliance of a building design with Standard 90.1.

ECB is used if a project does meet all of the prescriptive requirements in Standard 90.1 by trading off prescriptive requirements for energy cost reductions in other parts of the building design. Thus, an energy cost budget for the entire building is satisfied.

ECB accommodates unique designs and is more flexible than the other two available compliance methods.

LEED certification using ECB
- Show energy costs of the proposed design are less than or equal to the energy costs of a similar "budget" building

- Similar budget building meets minimum requirements of Standard 90.1
- Design building must use acceptable simulation software for modeling performance

ASHRAE STANDARD 189

ASHRAE Standard 189.1-2014: Standard for the Design of High-Performance Green Buildings (ANSI Approved; USGBC and IES Co-sponsored) except low-rise residential buildings.

Standard 189.1 provides total building sustainability guidance for designing, building, and operating high-performance green buildings.

- Site sustainability, energy and water use efficiency, indoor environmental quality (IEQ), and the building's impact on the environment.
- Compliance option of the 2012 International Green Construction Code™ (IgCC) that regulates construction of new and remodeled commercial buildings.

ASHRAE GREEN GUIDE

The Design, Construction, and Operation of Sustainable Buildings
- Guidelines suggestions to coordinate green projects amongst team members
- Information presented by project phase
- Reference on green-building design subjects

Contents Include:
- Green/sustainable high-performance design
- Commissioning
- Design process
- LEED guidance
- Engineering design-load, thermal comfort delivery systems
- Plumbing and fire protection systems
- Building control systems
- Documentation for construction
- Operation/maintenance/performance evaluation

SAMPLE QUESTIONS

1.   What is the primary function of ENERGY STAR's Portfolio Manager;
      and who operates the ENERGY STAR program?
      a.   Measure indoor air quality of green buildings; operated by US
            EPA
      b.   Rate high performing green buildings; operated by ASHRAE
      c.   Benchmark/measure/track building energy performance;
            operated by US Green Building Council
      d.   Understand the financial value of energy projects; operated by
            AEE
      e.   Benchmark/measure/track building energy performance;
            operated by US EPA and US Dept. of Energy

2.   Green Globes utilizes a point scale to assess the environmental impact
      of buildings in multiple categories of energy, indoor environment,
      site impact, water resources, emissions, and project/environmental
      management. The maximum points available on this scale is:
      a. 1
      b. 10
      c. 100
      d. 1,000
      e. 10,000

3.   The Leadership in Energy & Environmental Design (LEED) program
      administered by the U.S. Green Buildings Council (USGBC) includes
      five sub-programs. Which one of the following is not a LEED sub-
      program?
      a.   New Construction
      b.   Existing Buildings (Operations & Maintenance)
      c.   Commercial Interiors
      d.   Core & Shell
      e.   Military Facilities

4.   In version 3 of the LEED program, four of the rating systems (such as
      for EB-O&M) utilize a scale which includes base points, plus points
      for innovation and regional points. Under the auspices of this point
      system, what are the maximum total points which can be achieved?
      a.   75

    b.   100
    c.   110
    d.   1,000
    e.   10,000

5.   What is the minimum ENERGY STAR energy performance rating (from Portfolio Manager) required to meet the LEED EB-O&M prerequisite for the energy and atmosphere category?

    a.   60
    b.   69
    c.   75
    d.   80
    e.   100

SOLUTIONS TO SAMPLE QUESTIONS

    1.   e
    2.   d
    3.   e
    4.   c
    5.   b

Chapter 13

Thermal Energy Storage Systems

INTRODUCTION

Thermal storage is the temporary storage of energy at high or low temperatures for use at a later time when needed.

The time gap between when the energy is available and when it can be used may be anywhere from several hours to several months. Thus, operating cycles can be anywhere from daily to seasonal.

Topic Areas

Design Strategies	Operating Strategies
Storage Media	Advantages and Limitations
Chilled Water Storage	Ice Storage
Sizing	Volume Requirements
Full Storage Systems	Partial Storage Systems

DEFINITIONS

Sensible heat—the amount of heat that is added or removed from a substance or thermodynamic system that does not alter its state while its temperature changes.

Example: Sensible heat is added to water in a liquid state that increases its temperature from 75°F to 115°F, and the water remains in a liquid state.

Latent heat—the amount of heat that is added or removed from a substance or thermodynamic system during a change of state while its temperature does not change.

151

Example: Latent heat is added to a volume of water in a solid state, and its temperature remains 32°F while the water changes from a solid to a liquid state.

Energy storage may be in **sensible heat** or **latent heat**. The terms hot and cold refer to the temperature, not the *heat value*.
- Thermal storage at low temperatures is considered to be from 20°F to 70°F.
- Hot temperature storage is considered to be at temperatures of 70°F to 160°F.

DESIGN AND OPERATING STRATEGIES

Full Storage System—stored energy is used to meet an entire thermal load requirement during a period of time without assistance from thermal energy equipment.
- Stored cooled water is use in chiller system
- Stored hot water is used for heating or systems or domestic hot water use.

Load Shifting System—produces energy for storage during off-peak hours at lower energy rates for use during peak hours.

Partial Storage System—thermal energy equipment operates continuously and excess capacity during off peak time is stored and used to meet part of the peak period thermal load.
- Levels the load or limits demand
- Initial and operating costs of the equipment are minimized compared to full storage system
- Operates at a relatively constant capacity

STORAGE MEDIA

Materials used for thermal storage:
- **Water** is the most commonly used medium. Heated or cooled water is stored in tanks for later use.
- **Ice** is generated and stored for later use in cooling. The transfer of heat is accomplished using water or air.

- **Phase change materials** such as salt hydrates can be used to store the latent heat of fusion.
- **Subterranean rock beds** and **earth** are sometimes used for storage
  — Storage and containment of heat can be made with rock beds
  — Earth used as a source or heat sink for heat pumps
  — The temperatures involved with earthen storage makes it compatible with solar energy systems

CHILLED WATER STORAGE

Chilled Water Storage systems use sensible heat capacity of water to store cooling capacity.

Storage Tank size based mostly on temperature difference between supply water and return water
- Temperature differential of at least 16°F required
- A 20°F temperature differential is the practical maximum for most applications

ICE STORAGE

- **Ice** is generated and stored for later use in cooling
- **Heat transfer** is accomplished using water or air
- The high heat required for ice to liquid state change allows for **higher density energy storage for ice than for water**.
- A relatively low melting point of ice makes it useful for thermal storage in refrigeration use.

Methods
- **Ice harvesting**: Ice is formed on an evaporator surface and released to a partially filled water storage tank
- **External melt ice-on-coil**: Ice is formed on submerged pipes or tubes; the tubes carry a refrigerant or secondary fluid that is circulated
- **Internal melt ice-on-coil**: Ice is formed on submerged pipes or tubes, and warm coolant is circulated through pipes
- **Encapsulated ice**: Water in plastic containers reside in the tank; cold or warm coolant is circulated through the tank; the containers freeze and melt

- **Ice slurry**: Water in a water/glycol solution is frozen into a slurry, slushy mix (frozen water and liquid glycol) and pumped to a storage tank

SIZING AND VOLUME REQUIREMENTS

Thermal energy storage systems are designed to provide a cooling design **capacity**, and storage tanks are sized accordingly. Factors:
- Full Storage, Load Shifting, or Partial Storage System design impacts storage tank size and volume
- Tank volume is affected by maintaining separation of stored cold water and warm return water
- Tank storage volume calculation uses "Figure of Merit (FOM)"
  — measures a tank's ability to maintain separation of cold and warm waters
  — FOM = volume of water usable for cooling/total volume available
  — Well-designed stratified tanks typically have FOMs of 85 to 95 percent.
- Storage tank volume for chilled water is calculated using:
  — FOM
  — Total Ton-Hours
  — Tank Differential Temperature

ADVANTAGES AND LIMITATIONS

Savings from thermal storage can be realized in a number of ways:
- **May allow smaller and less expensive HVAC equipment** to be used. The operating costs of smaller capacity devices may also be less.
- **Alternative energy sources** such as solar and wind power may become feasible with the availability of storage.
- **Off-peak utility rates** may make the consumption of purchased energy at one time and use at another much cheaper.
- **Heat reclamation** may become easier with the use of storage.

Thermal storage becomes **a viable option** when there are:
- High loads for short periods of time

- High demand and peak rates imposed by utilities
- Capacity of building systems needs to be increased
  — Facility expansion
  — Installation of cogeneration

SAMPLE QUESTIONS

1. Which of the following criteria would never be used to determine the applicability of installing a TES cooling system?
   a. Cost per square foot of space
   b. Utility rate structure
   c. Capacity of the boilers
   d. Load profile of the building
   e. Availability of incentives from utilities

2. Which of the following rate structures would most lend itself to justifying a TES installation?
   a. $0.05/kWh, $5/kW-mo.
   b. $0.03/kWh, $10/kW-mo. Peak, $5/kW-mo Off peak.
   c. $0.08/kWh for all kWh < 1,000,000 kWh and $0.04 for all kWh >1,000,000 kWh.
   d. $0.03/kWh, $10/kW-mo. Peak, $2/kW-mo Off peak.

3. For an existing building installing TES, which of the following will most likely be favored?
   a. Load Leveling
   b. Load Shifting
   c. Ice Storage
   d. Ice Storage or chilled water.
   e. b and d

4. The only time a load leveling (partial shift) operating strategy would be used is in a retrofit situation when there is already a cooling system in place sized to handle the peak summer load.
   a. True
   b. False

5.   A 10-ton air-conditioner has an EER of 10.0. What is its electrical
      load?
      a.   8 kW
      b.   10 kW
      c.   12 kW
      d.   14 kW

SOLUTIONS TO SAMPLE PROBLEMS

          1.   c
          2.   d
          3.   e
          4.   b
          5.   c

Chapter 14

Lighting Systems

INTRODUCTION

While some topics on the CEM exam may not be considered in every energy assessment, lighting system evaluations are part of almost every assessment from residential to commercial to industrial (approximately 40% of energy load). Energy managers need to be familiar with the various terms related to lighting and their definitions. They also need to understand the benefits and drawbacks for each lighting technology and be able to calculate the energy consumption for various configurations.

Topic Areas

Light Sources	Efficiency and Efficacy
Lamp Life	Strike and Restrike
Lumens	Footcandles
Zonal Cavity Design Method	Inverse Square Law
Coefficient of Utilization	Room Cavity Ratio
Lamp Lumen Depreciation	Light Loss Factors
Dimming	Lighting Controls
Color Temperature	Color Rendering Index
Visual Comfort Factor	Reflectors
Ballasts	Ballast Factor
Lighting Retrofits	IES Lighting Standards
LED Lighting	

LIGHT SOURCES

Incandescent Lamp
    Characteristics:
*   Inexpensive, available in various shapes and wattages
*   Easily dimmed
*   Very inefficient

- Short lamp life
- Contribute to building heat
- Voltage-sensitive, with lamp life, lumen output, and wattage dependent on the applied voltage

Tungsten-Halogen (TH)

Halogen bulbs are a type of incandescent bulb. Characteristics:

- More efficient than standard incandescent lamps.
- The design increases the lumens per watt LPW (luminous efficacy).
- Provides a whiter light than standard incandescent bulbs.
- Offers a longer life than standard incandescent bulbs.
- Improved lamp lumen depreciation over incandescent bulbs.

Fluorescent Lamps

Fluorescent lamps are the most commonly used lamp type in commercial and industrial applications today. Light is generated when ultraviolet (UV) energy from a mercury arc strikes a fluorescent phosphor on the inside surface of a tube. Characteristics:

- Long life (12,000-20,000 hours)
- High efficacy (75-90 LPW)
- Excellent color rendering
- Temperature-sensitive
- Ballast required to both start and operate
- Least sensitive bulb to voltage variations

Compact Fluorescent Lamps (CFL)

CFLs are fluorescent bulbs with a folded or bridged tube design. Characteristics:

- Long life (10,000 hours)
- High color rendering index (CRI)
- Preheat and rapid-start models
- Alternatives to incandescent lighting
- Configurations include designs with two or more tubes

High-Intensity Discharge Lamps (HID)

These are classified as electric arc discharge lamps. The bulbs operate under high pressure and generate their light directly from an arc contained in a small tube enclosed in a larger outer bulb.

- HID lamps require time to strike
- Time required to cool down and restrike
- The National Electric Code (NEC) requires a backup lighting system for HID lighting for public safety
- High lumen ratings
- Long life

Included in this classification are:

— **Mercury Vapor (MV) Lamps**—oldest HID source and considered *obsolete*.

— **Metal Halide (MH) Lamps**
  - Arc tube contains additives called metal halides that provide a brighter, whiter light.
  - Improved lumen and color performance over MV lamps.
  - Note: for safety reasons, MH lamps must be turned off at least 15 minutes every week.
  - Group relamping before the end of rated life.
  - Uses a number of different metals, each of which give off a specific characteristic color:
    * dysprosium—broad blue-green
    * indium—narrow blue
    * lithium—narrow red
    * scandium—broad blue-green
    * sodium—narrow yellow
    * thallium—narrow green
    * tin—broad orange-red
  - Significant disadvantage is a tendency to shift colors as lamp ages

— **High Pressure Sodium (HPS) lamps**
  - Primary source for industrial lighting, highway, and street lighting
  - Characteristic yellow color, high efficacy
  - Life rating of 24,000 hours
  - Standard lamps cycle at the end of their life, indicating the need for replacement; advanced models do not cycle.
  - Double arc-tube HPS lamp used for safety and security applications.

LUMENS

The **lumen** is a unit of **luminous flux** that is a *measure of the total light from a source.*
Luminous Efficacy: Efficiency of light sources, calculated by dividing light output by the power input = lumens per watt (LPW). A higher LPW is a more efficient source.

ZONAL CAVITY DESIGN METHOD

The **Zonal Cavity Design Method**, also called the lumen method, is a process for determining the number of lamps needed in a specific area to provide adequate lighting. The lumen method uses the equation:

$$N = F \cdot A / (Lu \cdot L \cdot Cu) \qquad \text{where}$$

N is number of lamps needed
F is the required foot-candle level at the work area
A is area of the room square feet
Lu is lamp output in lumens
L is the depreciation factor for the lamp and fixture
Cu is coefficient of utilization

The **depreciation factor**, L, is the light delivered to the light emitted, after accounting for dirt and deposits in and on the lamp and fixture. The **coefficient of utilization**, Cu, is the ratio of delivered lumens to radiated lumens. Manufacturer's data are provided for a lamp's coefficient of utilization. If the photometric chart is provided, one can use the **room cavity ratio**, reflectance off the wall and reflectance off the ceiling to identify the Cu on the photometric chart.

Room Cavity Ratio (for regular rooms shaped like a square or rectangle)
= [5 x Room Cavity Depth x (Room Length + Room Width)] ÷ (Room Length x Room Width)

Room Cavity Ratio (for irregular-shaped rooms) = (2.5 x Room Cavity Depth x Perimeter) ÷ Area in Square Feet

Ceiling Cavity Ratio = [5 x Ceiling Cavity Depth x (Room Length x Room Width)] ÷ (Room Length x Room Width)

Floor Cavity Ratio = [5 x Floor Cavity Depth x (Room Length x Room Width)] ÷ Room Length x Room Width

Room Surface Reflectance (%) = Reflected Reading ÷ Incident Reading

Reflected Reading = Measurement from a light meter holding it about 1.5 feet away from the surface with the sensor parallel and facing the surface.

Incident Reading = Measurement from a light meter held flat against the surface and facing out into the room.

Calculating Number of Lamps and Fixtures and Spacing

Maximum Allowable Spacing Between Fixtures = Fixture Spacing Criteria x Mounting Height

Mounting height (MH): Distance in feet between the bottom of the fixture and the workplane

Spacing Between Fixtures = Square Root of (Area in Square Feet ÷ Required No. of Fixtures)

Number of Fixtures to be Placed in Each Row (N_{row}) = Room Length ÷ Spacing

Number of Fixtures to be Placed in Each Column (N_{column}) = Room Width ÷ Spacing

For the above two formulas, round results to the nearest whole integer.

Spacing row = Room Length ÷ (Number of Fixtures/Row - 1/3)

Spacing column = Room Width ÷ (Number of Fixtures/Column -1/3)

If the resulting number of fixtures does not equal the originally calculated number, calculate impact on the designed light level:

% Design Light Level = Actual No. of Fixtures ÷ Originally Calculated No. of Fixtures

To calculate fixtures mounted in continuous rows:

Number of Luminaires in a Continuous Row = (Room Length ÷ Fixture Length) - 1

Number of Continuous Rows = Total Number of Fixtures ÷ Fixtures Per Row

COEFFICIENT OF UTILIZATION (CU)

The **coefficient of utilization** expresses the efficiency of a luminaire in transferring luminous energy to the working plane. It is the ratio of luminous flux upon a work plane to that emitted by the bulbs inside the luminaire.

LAMP LUMEN DEPRECIATION (LLD)

Lamp lumen depreciation is also known as lumen maintenance. It is a measure of the percent of initial lumens and is one of several light loss factors used in lighting calculations. LLD can be calculated by dividing the mean (design) lumens by the initial lumen rating.

DIMMING

Dimming refers to reducing the amount of light output for a given luminaire. It may be accomplished by using a solid-state dimmer switch. The switch varies the amount of time that energy is transferred to the luminaire during a given cycle.

COLOR TEMPERATURE

The **color temperature** of a lamp is described in terms of its appearance (when lighted) to the eye.

- Defined using the Kelvin scale ranging from 1,500K, which appears red-orange, to 9,000K, which appears blue.
- Light sources with higher color temperature than 4,100K appear "cool" and those of a lower color temperature than 3,100K "warm."

VISUAL COMFORT PROBABILITY (VCP)

Visual comfort probability is a rating of lighting systems that is expressed as a percentage of people who will find the lighting system acceptable in terms of discomfort glare. IESNA minimum recommendation for office interiors is **70%**, and the recommendation for computer applications is **80%**.

BALLASTS

Ballasts are required to start and operate fluorescent and HID lamps (discharge lamps)

- Functions:
  — Provide the right voltage to start the arc discharge.
  — Regulate the lamp current to stabilize light output.
  — In rapid-start ballasts, a *third function* is to provide the energy to heat the electrodes.
- Original electromagnetic units that consist of a core of magnetic steel laminations surrounded by two copper or aluminum coils.
- Efficiency of magnetic ballasts has been improved by using low-loss magnetic material and copper windings, resulting in lower internal losses.
- New electronic ballasts.

Fluorescent ballasts are provided to operate fluorescent lamps in the following ways:

- Preheat: Lamp electrodes are heated prior to the application of a high starting voltage that initiates the arc discharge.

- The starting electrode voltage is applied through a "starter," a thermal switch that applies the high starting voltage across the electrodes.
- Lamps of less than 30 watts are usually operated preheat—causes "flickering" when starting.
- The ballast has two windings to provide the proper low voltage to the electrodes during starting and operation.
  — **Rapid-Start**: Lamp electrodes are heated prior to and during operation, characterized by smooth starting and long lamp life.
  — **Instant-Start**: Lamp electrodes are not heated, provide a high open-circuit voltage across the unheated electrodes to initiate the arc discharge. Instant-start operation is more efficient than rapid-start, but as in preheat, lamp life is shorter.
- Ballast efficiency must meet a minimum **ballast efficacy factor (BEF)** unrelated to efficiency.

BALLAST FACTOR

Ballast factor (BF) is a ratio:

$$= \frac{\text{Lamp's lumen output on commercial ballast}}{\text{Lamp's rated light output}}$$

The efficiency of fluorescent ballasts can be improved when fluorescent lamps are operated by an electronic ballast at high frequency—they convert the input power to light output more efficiently.
- The **LPW of the lamp/electronic ballast combination is increased**, producing more light for the same power, or producing the same light with lower power.
- System input watts can actually be less than the total of the lamp wattages; actual watt input is controlled by the ballast factor, which can range from 47 to 130 percent.

LUMINAIRES

Complete package of fixture plus lamp, ballast, lenses, etc. Optical system includes the lamp cavity and diffusing media and includes one or more of the following components:

- **Reflectors** redirect light; retrofit reflectors increase reflectance and performance.
- A **refractor** redirects light by refraction.
- **Lenses** alter directional characteristics of light that passes through.
- **Diffusers** scatter light below the ceiling plane in all directions to reduce brightness.
- **Louvers** use baffles to control light distribution and block light at certain angles.
- **Parabolic** luminaires reduce glare and control light levels.
- **Paracube** louvers are constructed as small plastic squares and have high visual comfort probability (VCP) but reduced lighting efficiency.

Types of Luminaires

Luminaires are classified according to the manner in which they control or distribute their light output. They can be direct (downward), indirect (upward) or direct-indirect.

Direct luminaires can be open or shielded.

Indirect luminaires radiate light up, to reflect off a ceiling.

Direct-indirect luminaires combine efficiency and high CU of direct luminaires with the light distribution characteristics of indirect luminaires.

Luminaire Light-Loss Factors

- **Luminaire surface depreciation** is the nonrecoverable light-loss factor that describes how light is lost as the reflecting surfaces degrade over time.
- **Luminaire dirt depreciation (LDD)** is the recoverable light loss factor that describes how light is lost from the initial illuminance provided by clean, new luminaires after dirt collects on the reflecting surfaces at the time when it is expected that cleaning will be done.

Photometrics

A **Photometric Report** for a luminaire shows the light distribution in the form of a polar graph and a table whose values represent the variation in candlepower of a luminaire in a given reference plane.

LIGHTING CONTROLS

- Switching
- Occupancy sensors
- Scheduling controls
- Photocells

Lighting Control Strategies

Tuning, daylight harvesting, and lumen depreciation compensation are the three major control strategies used to reduce lighting power.

- Tuning— adjusting the light output of luminaires to a specific level needed for a task or to achieve a particular aesthetic environment. (Dimming is one type of tuning strategy.)
- Daylight harvesting systems change the light level gradually according to the daylight level using photosensors. Unlike photocells that switch lights based on light level, these silicon sensors have an analog output that depends on the amount of light falling on them.
- Lumen depreciation compensation employs special photosensors that detect the actual light level and track the lumen depreciation of the lamps. When the lamps are new and surfaces clean, the output of the dimming ballasts and input power is low, saving energy. Input power and light level gradually increase as lamps age and surfaces accumulate dirt, to compensate for these depreciation effects.
- **Sensor Types**:
  — Ultrasonic personnel sensors offer the greatest coverage. Some of these sensors can cover up to 2,000 square feet.
  — A personnel sensor turns off or decreases the lighting automatically when there are no people in the covered area.
  — Some audio sensors can cover up to 1,600 square feet.
  — Infrared sensors cover much less area, up to 130 square feet.

LIGHTING EFFICIENCY AND RETROFITS—STEPS

1. Use the most efficient lamp-ballast combination. Lamps and ballasts should be selected as a system and the ballast factor (BF)

taken into consideration.

2.   Use the most efficient luminaire for the application.
3.   Consider increasing the efficiency of the space with light-colored ceilings, walls, and partitions.
4.   Consider task-ambient lighting.
5.   Consider lighting controls to limit the time lights are on in unoccupied spaces, or to reduce lighting power.

MAINTENANCE

- The light output of lighting systems decreases over time
- Many lighting systems are overdesigned to compensate for future, unnecessary light loss
- Improving maintenance practices can reduce light loss and allow reductions in energy consumption or improve light levels.
- Group maintenance practices save money.
- Proper maintenance is the most neglected, most cost-effective way of reducing the overall cost of lighting.
- When maintenance is not performed, performance suffers gradually. The final result is a degraded lighting system performing at as low as 50 percent of its capability.
- Lumen depreciation, luminaire dirt depreciation, and room surface dirt depreciation light loss factors can be recovered by performing maintenance.

MEASUREMENTS

- **Luminous Efficacy** = Lumens/Watts
- **Color Temperature**: Appearance of light generated vs. black bodied radiator at that temp on the absolute temp scale (degrees Kelvin).
- **Color Rendering Index (from 0 to 1)**: How well can you distinguish colors under a light with a particular color temperature.
- **Illuminance**: Light level; the amount of light produced by a light source. Equal to luminous intensity divided by the square of the distance between the light source and the point of observation.

Measurement Devices
   Light measurement
   Digital Illuminance Meter
   Light Meter or Foot-candle meter measures **light intensity,** the **amount of illumination** the inside surface of a one-foot-radius sphere would be receiving if there were a uniform point source of one candela in the exact center of the sphere. Alternatively, it can be defined as **the illuminance on a** *one-square-foot* **surface** of which there is a uniformly distributed flux of one *lumen.*

Other Devices
   Operating Hour Monitor
   Occupancy monitor
   Occupancy sensor

Lamp Maintenance
   Calendar Lamp Life (Years) = Rated Lamp Life (Hours) ÷ Annual Hours of Operation (Hours/Year)

   Lamp Burnout Factor = [1—Percentage of Lamps Allowed to Fail Without Being Replaced]

Lighting Survey
- Location
- Lighting level
- Hours and Days of operation
- # fixtures
- # lamps on
- Color
- Task Lighting
- Required light levels

Regulatory
   Proper Disposal
   Energy Act 1992—restricted production of incandescents, T12s; established lighting efficacy and color index standards

Energy Efficiency Retrofit Opportunities
- Electronic Ballasts to replace coil or core ballasts
- Lighting Controls: dimmers, occupancy sensors
- Access to switches in occupied rooms

- Timers
- Photocells to detect daylight, turn off lighting

Series resistors are used as ballasts to control the current through LEDs.

Top 10 Lighting Retrofits

10. Occupancy sensors for interior lighting control. Choose the right technology and ensure it is applied correctly. Remember, passive infrared sensors (PIR) cannot see behind obstructions.

9. Astronomical electronic timer for exterior lighting control. This means lights on at dusk, off at dawn; an additional channel allows timed control of lights at fixed times. It is a cost-effective alternative to photocells.

8. Switch/timer for interior spaces where sensors are not cost-effective. Use the preset, electronic switchplate or switch version, which is low-cost and has a good payback. Use switch-type for electronic ballasts; the warning flick model is recommended.

7. Exit light LED (solid-state light-emitting diode) kit. Either LED tubes or arrays; products that overdrive the LED's will shorten life and cause lamp lumen depreciation.

6. T8/electronic ballast fixture upgrade. Retrofit or change to new fixtures, depending on cost-effectiveness, the need to retain UL listing, and ceiling material (asbestos). New fixtures optimize performance and retain UL listing. Retrofits are good for asbestos ceilings, but violate UL listing if sockets are relocated.

5. Exit light retrofit lamps. Try special long-life incandescent lamps, usually in a "twist-of-the-wrist" (TOW) light tube.

4. T10 "quick and dirty" fixture upgrade. The simplest is a four-lamp to two-lamp retrofit using two 3,700-lumen T10 lamps and one of two existing T12 two-lamp ballasts (remove and dispose of the second ballast). This works well for 34-watt T12 replacements; disadvantage is T10 cost.

3. Self-ballasted compact fluorescent lamps (CFLs). In a TOW retrofit to replace incandescent lamps, watch equivalency recommendations; you usually need one size higher because of multiple sensitivities of CFLs that need to be calculated for actual lumen output (position, temperature, BF, and LLD).

2. R and PAR lamp replacements. Retrofit to replace older, less-ef-

ficient incandescent floods and spots. Newer bulbs replace those banned by EPAct and have the same light output and beam patterns. Usually tungsten- halogen (T-H), krypton R lamps, or ER lamps replace 75R30 in down-lights.

1.   Energy-efficient incandescent bulbs. A retrofit using reduced wattage bulbs with the same light output as standard wattage incandescent bulbs. Krypton gas is an older technology, T-H is the newest.

Common Lamp Characteristics and Definitions

Average Rated Life: This is the median value of life expectancy assigned to a lamp, in hours, at which half of a large group of lamps have failed.

Color Rendering Index (CRI): Color rendition describes the effect a light source has on the appearance of colored objects. The color-rendering capability of a lamp is measured as the CRI.
- The higher the CRI, the less distortion of the object's color by the lamp's light output.
- The scale ranges from 0 to 100. A CRI of 100 indicates that there is no color shift as compared to a reference source. The lower the CRI, the more pronounced the shift may be. CRI values should only be compared among lamps of similar color temperature.

Quantity of Light

The total light from a light source is measured in **lumens**. Lamps are now labeled with measured lumen ratings and efficiency ratings (efficacy).
- The quantity of light that falls on a work surface is measured in **footcandles**. One **footcandle = one lumen per square foot of area**.
- An **illuminance** meter is a useful tool to measure the amount of light in work spaces. It is important to understand that the footcandle measure indicates only a level of illuminance.

Quality of Light
- Lighting quality is complex and there is no single measure
- highly subjective and not easily quantified
- certain quality *characteristics*

Veiling reflections (reflected glare) detract from lighting quality by obscuring task details by reducing contrast—most noticeable from luminaires located in front of and above the viewing task.

Glare: is defined as the sensation produced by brightness within the visual field, sufficiently greater than the brightness to which the eye is adapted to cause a loss of visual performance. Building interiors has *discomfort glare*.

Remove glare:
- Louvers such as are found in deep-cell parabolic luminaires or low-glare acrylic lenses that reduce surface brightness at high viewing angles
- Use of indirect lighting

Lamp color also affects lighting quality. Recommendations regarding pleasant combinations of lamp color temperature and illuminance are changing and are best left to building occupant preferences.
- When lamps of good color rendering are used, illuminance may be lowered to achieve equivalent brightness and visual clarity. Example: When upgrading from cool-white lamps to higher CRI T8 lamps, employees may respond that the new lamps are "too bright."

Lamp flicker can also reduce lighting quality.
- Flicker is especially noticeable at high light levels, such as industrial inspection lighting.
- Electronic ballasts that operate fluorescent lamps at high frequency can reduce flicker to an imperceptible level.

To improve lighting quality, it is important to **balance office lighting for visual performance and for visual comfort**.

SAMPLE QUESTIONS

1. A room with T12 cool white lamps is retrofitted with T8 tri-phosphor lamps. The changes produced in this retrofit will be:
    a. Slower lumen depreciation and higher lumens per watt for the lamps
    b. Higher initial lumen output of the lamps and a lower CRI
    c. Answer a and higher initial lumen output
    d. a and b

e.　None of the above

2.　An office space lighting system is determined by the Lumen Method to require 65 F40T10 lamps. The following are the assumptions made in the evaluation:

The required footcandle level at the task, is 70 footcandles; each lamp puts out 3,000 lumens; the product of the lamp and dirt depreciation factors is 0.6; and the coefficient of utilization is 0.6. The maximum area of the office, in sq. ft, would be about:

a.　500 ft^2
b.　1,000 ft^2
c.　1,500 ft^2
d.　2,000 ft^2

3.　A room, with a wall reflectance of 30% and a ceiling reflectance of 70%, is lighted by 100 lamps. The floor is square with sides of 50 feet and the height of the lamps above the work surface is 10 feet. The room is now painted and the wall reflectance goes up to 70%. Using the lumen method, approximately how many lamps can be removed?

a.　2 lamps
b.　5 lamps
c.　7 lamps
d.　9 lamps

4.　Ten F40T12 40-watt 2-lamp fixtures with conventional magnetic ballasts are to be retrofitted with F40T8 32-watt lamps with electronic ballasts. The 40-watt fixture consumes 96 watts and the proposed fixture will consume 62 watts. The cost of the retrofit is expected to be $38/fixture, which includes the cost for ballast, lamp and sockets. If electricity costs 7 cents/kWh and $8/kW/mo and the lights are "on" 3,000 hours/yr, and, the project lasts 5 years, what is the rate of return of this investment?

a.　Between 11% and 12%
b.　Between 14% and 15%
c.　Between 19% and 20%
d.　None of the above

5.　Five F40T12 40-watt 2-lamp fixtures are instead fitted with 34-watt

lamps. If ballast consumption is 15% of the total electrical input to the lamps in both cases, and the annual operating hours are 2,900, what are the annual kWh savings?
a. 100 kWh
b. 150 kWh
c. 175 kWh
d. 200 kWh

SOLUTIONS TO SAMPLE QUESTIONS

1. a
2. b
3. d
4. a
5. d

Chapter 15

Boiler and Steam Systems

INTRODUCTION

Boilers and steam systems are used for space heating in commercial and residential buildings, as well as for industrial processes, using the basic concepts first included in early designs used 100+ years ago.

At its most basic level, a *boiler* is a chamber in which steam or hot water is produced for space heating throughout a facility, or for an industrial application.

This chapter covers the operational aspects of boiler and steam systems, including the combustion of fuel, system components, and energy efficiency.

Topic Areas

Combustion Efficiency	Boiler Economizers
Air to Fuel Ratio	Waste Heat Recovery
Excess Air	Steam Traps
Condensate Return	Steam Leaks
Boiler Blowdown	Turbulators
Flash Steam	HHV and LHV
Scaling and Fouling	Condensing Boilers

BOILER SYSTEM OVERVIEW

A **boiler** is a chamber that holds water to be heated to a desired temperature and pressure. Heated water in the form of steam (above 212°F) or hot water (below 212°F) is used as the medium to carry heat to distributed points throughout a facility.

Boiler systems using steam are, naturally, called *steam* systems; boilers that utilize hot water are known as *hydronic* systems. Tubes inside the boiler are utilized based on the boiler type. Two of the most common boiler types are:

- **Fire Tube Boiler**—heat in boiler tubes transfers to water surrounding the tubes. The largest number of boilers installed in buildings and factories are fire tube boilers.

- **Water Tube Boiler**—heat in the boiler surrounding tubes transfers to water in the tubes.

Space heating uses the following process:
- A fuel is *combusted* (burned) to generate heat
- Heat is transferred to water in the boiler to produce either steam (vapor) or higher temperature water
- Steam or hot water flows through distribution pipes to *convectors* (often called *radiators*) at multiple points within a building
- For space heating, convectors transfer heat from the steam or hot water to the surrounding air
- Steam *condensate* (condensed steam from change of state in radiators) or reduced temperature water from convectors is returned to the boiler and the process is repeated

A **burner** combusts fuels such as oil or natural gas and generates the heat applied inside a boiler. Figure 15-1 shows a boiler system with burner in foreground.

Figures 15-2 shows tubes inside a boiler.

Steam Pipe Distribution

A steam distribution and condensate return configuration is typically a one-pipe or two-pipe system. A one-pipe system utilizes one pipe for both steam distribution to convectors (radiators) and condensate return to the boiler. A two-pipe system utilizes one pipe for steam distribution to convectors (radiators) and a second pipe for returning condensate from the radiator to the boiler. Figure 15-3 illustrates a one-pipe boiler steam distribution system and Figure 15-4 a two-pipe system.

A **convector** is a device that transfers thermal energy from one place to another by the movement of fluids.
- Heat from steam or hot water enters its core and is transferred to the surrounding space.
- Made of materials such as steel, aluminum, or cast iron to effectively transfer heat.
- Air circulation using fans may be added to disperse heat to a space.
- Post heat-transfer, water is returned to the boiler for re-heating.

Figure 15-1. Boiler system with burner in foreground.

Figure 15-2. Boiler tubes.

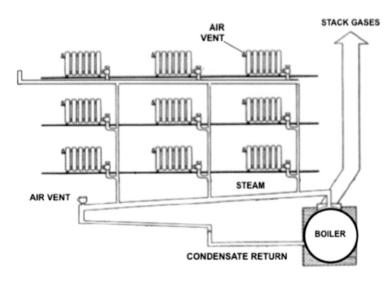

Figure 15-3. One-Pipe Steam Distribution System

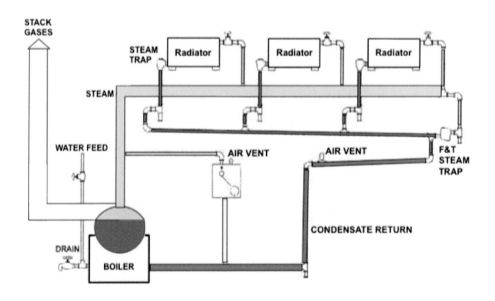

Figure 15-4. Two-pipe Steam Distribution System

A **radiator** is a common term for a convector.

Boiler Efficiency

$$\text{Boiler Efficiency \%} = \frac{\text{Heat Output (Btus)}}{\text{Energy Input (Btus)}} \times 100$$

Typical steam boiler efficiency = 75% to 85%.
Typical hot water boiler efficiency = 85% to 95%.

The Combustion Process

Combustion is the process in which Oxygen (O_2) in air reacts with the combustible components of a fuel, resulting in the generation of heat and chemical compound gases.

- Combustion takes place in the fuel burner.
- Fuels consist mostly of atomic Carbon (C), Hydrogen (H), Oxygen (O), Nitrogen (N), Sulfur (S), minerals (ash) and water (H_2O).
- Combustion generates heat by the chemical reaction of Oxygen to produce CO_2, CO, NO_x, C_nH_m, SO_2, SO_3, and H_2O, which are all environmentally harmful gases except for H_2O.
- Generation of CO is an incomplete combustion reaction in which CO_2 ideally would have been generated.

Gases are exhausted to the atmosphere from a boiler through the chimney flue referred to as the stack. Gases generated by combustion are called stack gases or flue gases.

Improving boiler efficiency, which includes combustion efficiency, reduces the environmental and health threats of stack gas emissions.

COMBUSTION EFFICIENCY

Combustion efficiency considers:
- Characteristics of the fuel
- Stack temperature (temperature of the stack gases)
- Heat losses of stack gases, CO production, and moisture

The theoretical minimum amount of air for combustion is called **stoichiometric air**.

- The stoichiometric air would completely combust fuel to carbon dioxide (CO_2), water (H_2O), and sulfur dioxide (SO_2) if sulfur is present.
- Combustion converts carbon (C) in the fuel to a maximum amount of CO_2.
- For complete combustion to occur, an amount of air is supplied in excess of what is theoretically required due to less than ideal mixing of fuel and air during the combustion process.
- Excess air helps achieve complete conversion of carbon to CO_2, and minimizes CO production.

AIR-TO-FUEL RATIO

Air-to-fuel ratio is the amount of stoichiometric air required to completely combust one mass unit of fuel. Using kg for mass:

$$\text{Air-to-Fuel Ratio (AF)} = \frac{\text{stoichiometric air mass (kg) for combustion}}{\text{fuel mass (1 kg)}}$$

- The AF is calculated from the ultimate chemical composition of the fuel.
- The AF ratio has nothing to do with the boiler system design or the combustion process of the system.

Table 15-1 lists air to fuel ratios for five fuels.

Table 15-1. Air-to-fuel ratio of fuels

Fuel	Phase	AF
Light fuel oil	liquid	14.1
Medium heavy fuel oil	liquid	13.8
Heavy fuel oil	liquid	13.5
Gasoline	liquid	14.7
Natural Gas	gas	17.2

EXCESS AIR

Excess air ratio, or *factor*, is the amount of air mass required to combust one mass unit of fuel, divided by the air to fuel ratio. Using kg for mass:

$$\text{Excess Air (EA)} = \frac{\text{Air mass (kg) for combustion of fuel mass (1 kg)}}{AF}$$

Examples:

- A burner requiring 20% excess air to correctly combust fuel oil is has an excess air factor of 1.2.
- A combustion process requiring 100% excess air has an excess air factor of 2.

The theoretical, ideal combustion process would require 0% excess air, and therefore have an excess air factor of 1.

The following excess air factors are achieved with careful monitoring and regular adjustment of combustion air at varying loads:

- Gas burners, forced draft   1.1 - 1.3
- Atmospheric gas burners   1.25 - 1.5
- Oil burners   1.15 - 1.3
- Coal dust burners   1.2 - 1.3
- Coal firing (mechanical)   1.3 - 1.5
- Coal firing (hand)   1.5 - 2.5

Average excess air factors of system operation may be much higher.

HEATING VALUE OF A FUEL: HHV AND LHV

The **heating value** or heating content of a fuel is the amount of heat released when combusted. Fuel heat value can be expressed in two ways:

- **Higher Heating Value (HHV)** is the heat released by a specified quantity of combusted fuel including the latent heat content of vaporization of water.
- **Lower Heating Value (LHV)** is the heat released by a specified quantity of combusted fuel not including latent heat content of vaporization of water.
- For HHV and LHV:
  - The quantity of fuel is at a pre-combusted set temperature, and heat content is measured after all combusted elements are brought back to the original pre-combustion set temperature.

— Heat content of water vapor is approximately 10% of the total heat content.
• LHV values are approximately 10% less than the HHV value for a given fuel.

WASTE HEAT RECOVERY AND BOILER ECONOMIZERS

The combustion process produces stack gases containing heat lost to the atmosphere. **Waste Heat Recovery** is the process to capture lost heat for useful work.

A **boiler economizer** is a unit that captures heat from the stack and returns it to the boiler water or other water system.

CONDENSING BOILERS

A **condensing boiler** is a hydronic boiler that supplements the conventional combustion process for heat generation with energy obtained by condensing water vapor contained in stack gases. The condensing process recovers heat from the stack for productive use.

A heat exchanger transfers heat from the vapor to the boiler by change of state from vapor to liquid, and the liquid is then discarded (drained).

A condensing boiler can achieve a rated Annual Fuel Utilization Efficiency (AFUE) of 95%.

SCALING AND FOULING

The process in which boiler water evaporates and causes chemical deposits to form on surfaces is called **scaling**. The chemical deposits called scale are mainly due to calcium and magnesium salts (carbonates or sulphates). Deposits may also include, phosphate, iron and silicon dioxide (silica).

Fouling is the accumulation of scale that impacts the ability to transfer heat within a boiler. Fouling may lead to tube overheating and failure.

Proper boiler maintenance includes water treatment to remove scale, including:

- Sodium carbonate
- Sodium hydroxide
- Sodium phosphate
- Organic polymers
- Starches

CONDENSATE RETURN

Condensate Return is the condensed steam in liquid form that returns to the boiler for re-heating and repeated circulation. Proper water levels in the boiler system must be maintained to provide proper heating distribution; effective condensate return minimizes the amount of feed water, also called *make-up water,* to be added to the system.

Make-up water should be de-mineralized to reduce scaling and fouling.

BOILER BLOWDOWN

Boiler blowdown is the process of removing water from the boiler that contains impurities as a result of steam evaporation. Steam pressure is utilized to blow water out of the boiler, and make-up water is then added.

FLASH STEAM

Flash steam is created by release of hot water condensate from high to low pressure within a steam system. Flash steam can be produced:
- During boiler blowdown
- At steam traps

STEAM TRAPS

A **steam trap** is a device used to remove condensate and remaining gases from convectors and distribution lines while retaining steam. Most steam traps are essentially automatic valves.

Functions
- Condensate removal that forms in steam lines
- Condensate and air removal from a convector or heat exchanger
- Increase temperature of a product using jacketed pipes or tubing (*steam tracing*)
- Prevent water hammer (banging pipes)
- Improve steam quality for other processes

Thermostatic Traps—use temperature to open and close a valve.
Designs
- Bellows trap—uses temperature of a fluid inside a chamber to open and close a valve.
- Bimetallic trap—uses thermal expansion of metals to open and close a valve.
- Thermal expansion trap—uses a thermostatic element filled with oil that heats and cools to move a piston and close/open a valve.

Mechanical Traps—use the difference in density between condensate and live steam to produce a change in the position of a float or bucket to open or close a valve.
Designs
- Ball float trap—relies on movement of a ball to open and close the outlet opening in the trap body
- Float and lever trap—uses ball float connected to a lever to change valve position
- Float and thermostatic traps (F&T) trap—float and lever trap with a thermostatic element to discharge air upon start-up
- Inverted bucket trap—uses entering steam pressure to move the inverted bucket upward and close a valve to retain the steam
- Open bucket trap—uses condensate to float the open bucket upward and close a valve.

Thermodynamic Traps—use difference in either kinetic energy or velocity between condensate and live steam to open and close a valve.
Designs
- Disk trap—uses pressure differences in steam velocities below and above a flat disk to move the disk upward and downward to control steam and condensate flow.

- Piston trap—uses flashing steam and the associated change in pressure to force a valve closed. Also called an *impulse trap*.

Applications and Steam Trap Options
- Condensate removal (drip applications)—condensate removal that forms in steam lines when heat is transferred from the steam and radiates to the space.
  — Relatively small condensate capacities
  — Minimal air volume discharge necessary
  — Use thermodynamic traps for line pressures over 30 psig
  — Use float & thermostatic for line pressures up to 30 psig
  — Inverted bucket traps used to handle large amounts of dirt and scale
- Condensate and Air Removal (process applications)—condensate and air removal from a convector or heat exchanger during a specific heat transfer process. Includes convectors (radiators) in two-pipe steam systems
  — Require larger condensate handling capability
  — Large air volume discharge necessary
  — Use Float & thermostatic, and thermostatic traps
- Increase temperature of a pipeline using jacketed pipes or tubing (steam tracing)
  — Wrapping a pipeline with steam tubing to heat the pipeline substance
  — Typical application is to raise viscosity of oil to improve flow through the pipeline
  — Thermostatic and bi-metallic traps used to remove condensate

STEAM LEAKS

Steam leaks reduce the efficiency of a boiler system. Steam leaks reduce system pressure and force the system to produce additional steam to make-up for losses.

Costs of steam losses are calculated using the formulas in the Appendix.

Factors to calculate steam losses:
- Steam operating pressure
- Ambient outside temperature

- Boiler efficiency
- Fuel cost
- Area of the orifice in which steam is lost

TURBULATORS

A **turbulator** is a device inserted within boiler tubes to improve heat transfer and therefore efficiency. The device, in the form of a twisted metal band, alters a constant flow of heat to be more turbulent and improve heat transfer to water.

SAMPLE QUESTIONS

1.   Steam traps are effective:
      a.   Non condensable gas releasing devices
      b.   Pressure differential devices
      c.   Condensate release devices
      d.   All of the above
      e.   None of the above

2.   The most effective method to constantly monitor boiler efficiency is by using a:
      a. O_2 monitor
      b.   CO monitor
      c.   Flue gas temperature monitor
      d.   Fuel flow gauge
      e. NO_X monitor

3.   How much energy is required every hour to heat 100 pounds per hour of 70 °F water to 300 °F saturated steam?
      a.   23,155 Btu/h
      b.   56,740 Btu/h
      c.   95,450 Btu/h
      d.   114,166 Btu/h
      e.   176,512 Btu/h

4.   What is the combustion efficiency of a natural gas fired boiler if

the carbon dioxide level in the flue gas is 9% and the stack gas temperature rise is 300 °F?

a. 84%
b. 82%
c. 80%
d. 78%
e. 76%

5. A boiler has 500 pounds per hour of blowdown at 60 psia which is sent to a pressurized recovery tank at 30 psia. How much 30 psia steam can be recovered for use in the facility?

a. 8.4 lb/h
b. 22.9 lb/h
c. 45.8 lb/h
d. 54.9 lb/h
e. 69.0 lb/h

SOLUTIONS TO SAMPLE QUESTIONS

1. d
2. b
3. d
4. a
5. b

Chapter 16

Maintenance and Commissioning

INTRODUCTION

Maintenance and commissioning are grouped together since both involve the proper application of skilled labor to ensure that building systems are working correctly. For many maintenance topics, the relevant equations are included in the reference section. This section complements the previous sections with additional details. Commissioning refers to a set of activities taken to ensure that equipment is configured correctly, properly applied and maintained to meet the design intent.

Topic Areas

Combustion Control	Compressed Air Leaks
Steam Leaks	Steam Traps
Insulation	Outside Air Ventilation
Group Relamping	Scheduled Maintenance
Preventative Maintenance	Proactive Maintenance
Boiler Scale	Water Treatment

COMBUSTION CONTROL

Effective **combustion** requires the proper balance between fuel and air.

Figure 16-1 illustrates the relationship between percent excess air and combustion efficiency as a function of stack temperature rise. If adjustments are made to the boiler fuel/air ratio to improve operating efficiency, Figure 16-1 will demonstrate the change in efficiency and the fuel savings can be calculated using:

$$\% \text{ Fuel Savings} = (\text{Eff}_{new} - \text{Eff}_{old}) / \text{Eff}_{new}$$

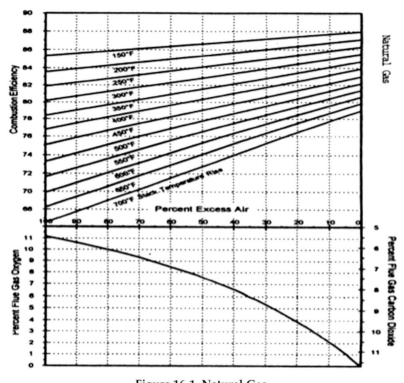

Figure 16-1. Natural Gas

The money savings are then calculated using:

Savings = % Fuel Savings (in decimal form) * Fuel Consumption

STEAM LEAKS

Figure 16-2.

INSULATION

Insulation serves the purpose of providing an isolation barrier between two different temperature surfaces. It can reduce uncontrolled energy losses and protect equipment and personnel. For steam lines, the difference in temperature between the steam and the outside is very large which creates a huge energy loss if the pipes are not insulated. Figure 16-3 demonstrates the potential losses due to uninsulated steam lines.

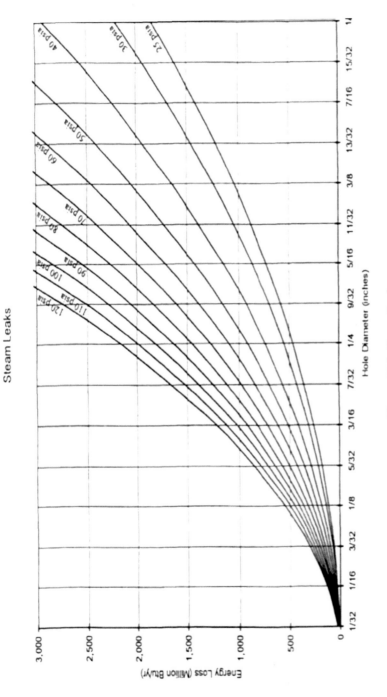

Figure 16-2. Steam Leaks

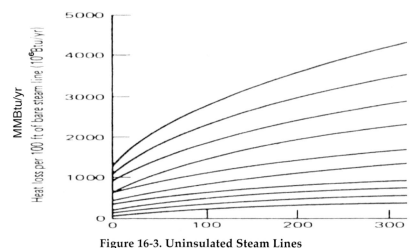

Figure 16-3. Uninsulated Steam Lines

GROUP RELAMPING

Some facilities save money by changing out all of the lights in a space at one time when they have reached 70% of their rated life instead of changing out lights one at a time as they burn out. The benefits of this approach include lower maintenance costs, more light, fewer unreplaced burnouts, less lamp stocking and fewer work interruptions.

The **group relamping interval** (GRI) is a function of the average rated lamp life and the annual fixture operation.

$$GRI = (Average\ Rated\ Lamp\ Life)\ *$$
$$0.7/(Annual\ Fixture\ Operation\ Hours)$$

PREVENTATIVE MAINTENANCE

Preventative maintenance refers to the category of actions that firms may take to keep equipment running to manufacturer specifications. Actions are taken on a set schedule. This includes establishing programs to regularly test for compressed air leaks, steam leaks, stuck steam traps, etc. after certain time intervals have passed.

COMPRESSED AIR LEAKS

One of the challenges associated with operating a compressed air system is how common air leaks will show up in a system. Plant managers may not realize the cost associated with air leaks. Figure 16-4 illustrates the relationship between air leaks and energy lost.

To solve for the standard cubic feet per minute (SCFM) lost due to an air leak:

$$\text{Leakage (SCFM)} = V * \frac{(P_1 - P_2)}{14.7 \text{ lb/in}^2 \, (T)}$$

V = total system volume (ft³)
P_1 = initial pressure (psig)
P_2 = final pressure (psig)
T = time in minutes

Costs of Compressed Air Leaks

Hole Diameter (in)	Energy Loss at Pressure (kWh/year)		
	110 psi	100 psi	90 psi
3/8	226,100	208,100	190,000
1/4	100,500	92,500	86,300
1/8	25,100	23,100	21,100
1/16	6,300	5,800	5,300
1/32	1,600	1,400	1,300

Figure 16-4. Compressed Air Leaks

STEAM TRAP MAINTENANCE

Most steam traps in industry are mechanical devices that are designed to extract condensate from steam lines and to bleed off air. Unfortunately, if they are not properly maintained they can become stuck in one position. If stuck open, they act like a steam leak in the system. Figure 16-5 demonstrates the relationship between steam leaks and energy lost.

OUTSIDE AIR VENTILATION

To maintain proper air quality in a facility, outside air must be brought in while stale inside air is exhausted. Louvers that operate to control the flow of outside air have mechanical linkages that need regular maintenance.

SCHEDULED MAINTENANCE

Scheduled maintenance refers to actions that are scheduled to occur after set time intervals to ensure equipment is operating to manufacturer specifications.

PROACTIVE MAINTENANCE

Proactive maintenance consists of two parts: preventative maintenance and predictive maintenance. It is a preventative maintenance strategy that allows a company to schedule production shutdowns for repairs and maintenance. Typical preventative maintenance is done based upon periodic time intervals. Predictive maintenance techniques allow firms to evaluate the existing system status such that one can predict when maintenance should be performed.

WATER TREATMENT

Water treatment in heat exchangers and boiler systems is important since impurities in makeup water can decrease heat transfer efficiency. If water ph is not maintained correctly, it can cause scale buildup on the heat transfer surfaces. Scale acts like insulation and blocks effective heat transfer. Figure 16-6 illustrates the costs associated with scale buildup in a boiler.

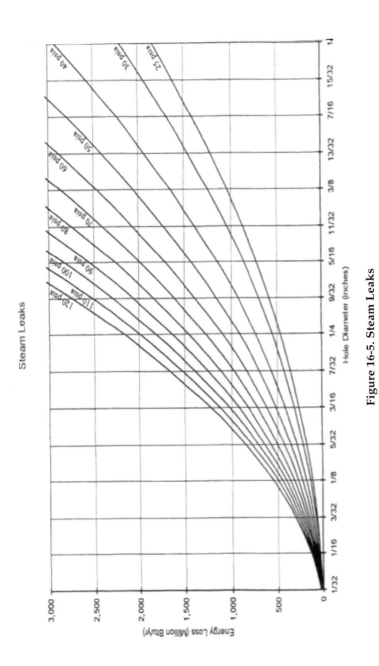

Figure 16-5. Steam Leaks

ENERGY LOSS FROM SCALE DEPOSITS

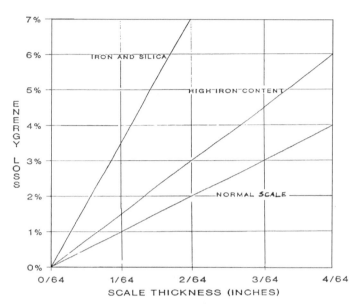

Figure 16-6

SAMPLE QUESTIONS

1.   Most steam traps rarely fail in the open position, so annual
      maintenance is cost-effective for locating and fixing the few traps
      that have failed open.
      a.   True
      b.   False

2.   A heavy commercial facility has a 100 psig compressed air system
      with two 1/16-inch leaks and three 1/8-inch leaks. The air line is
      energized 5000 hours per year. Electric energy costs $0.055/kWh.
      What is the annual cost of the air leaks?
      a. $8900/yr
      b. $5000/yr
      c. $4450/yr
      d. $2540/yr
      e. $1995/yr

3. A manufacturing plant has a 200 psig steam system, and a 40 foot section of 3-inch steam pipe that is not insulated. The steam system is operated 3500 hours a year, and the boiler efficiency is 70%. If boiler fuel costs $4.50 per million Btu, what is the annual cost of the uninsulated pipe (± $250)?
    a. Around $6500/yr
    b. Around $2500/yr
    c. Around $1250/yr
    d. Around $500/yr
    e. No answer is anywhere close

4. Predictive maintenance like vibration analysis, alignment and oil analysis can have the following impacts:
    a. Reduced unplanned failures
    b. Improved efficiencies
    c. Reduced repair costs
    d. All of the above
    e. None of the above

5. A company has proposed to conduct a complete steam leak test and will repair all identified leaks for a cost of $20,000. Your system operates 8760 h/yr at 80% efficiency, gas costs $4/MMBtu, and your steam gauge reads 95 psig. How many ¼" steam leaks will they need to find to equate to a 1-year payback?
    a. 1
    b. 2
    c. 5
    d. 10
    e. 20

SOLUTIONS TO SAMPLE PROBLEMS

    1. b
    2. d
    3. c
    4. d
    5. b

Chapter 17

Energy Savings Performance Contracting and Measurement and Verification

Energy efficiency projects are funded from a variety of sources including internal operating and capital expense budgets, as well as external financing programs. Energy Service Companies (ESCOs) and financing firms enter into performance contracts that reduce or eliminate upfront capital expenses based on their breadth of experience implementing energy efficiency projects that have resulted in lower operating costs.

Performance contract energy savings must be properly verified to satisfy contractual commitments and provide profit to ESCOs. An overview of performance contracting, measurement and verification (M&V) methodologies used to confirm energy cost reductions, and financing vehicles are presented in this chapter.

Topic Areas

Measurement and Verification Protocols	Energy Service Companies
Energy Savings Performance Contracting	Utility Financing
Shared Savings Contracts	Demand-side Management
Contracting and Leasing	Savings Determination
Risk Assessment	Loans, Stocks and Bonds

MEASUREMENT AND VERIFICATION PROTOCOLS AND SAVINGS DETERMINATION

Measurement and Verification (M&V) is the process to determine the actual energy savings due to the implementation of an **Energy Cost Measure (ECM)**.

- An M&V plan is developed and finalized after all stakeholders in the process (purchaser, vendors, service providers, etc.) agree on content and approach.
- **Measurements** are required, rather than energy use estimates, to be considered an M&V process.
- Energy that is no longer consumed following an ECM implementation, that is the *"absence of energy"* previously required, cannot be measured; therefore, a comparison must be made of a measured pre-implementation baseline to post-installation measured energy use.
- M&V components include:
  — Meter installation and calibration
  — Data collection
  — Analysis and statistical calculations of measured data
  — Reporting
  — Quality assurance
  — Third-party verification of reports

M&V Purpose

M&V can be used by facility staff, energy service companies (ESCOs), project investors, and financing firms for the following purposes:
- Verify energy savings
- Increase energy savings
- Support performance-based energy savings contracts
- Establish credibility for additional project financing
- Improve facility operations and maintenance
- Emission-reduction compliance
- Confirm effectiveness of state, city, and utility energy efficiency programs

Protocol Standard

The *International Performance Measurement and Verification Protocol (IPMVP)* is an industry standard, offered by the Efficiency Valuation Organization.

IPMVP Options to Determine Savings

Option A. Retrofit Isolation: Key Parameter Measurement
- Savings determined by field measurement of *one or more key performance parameters* that define the energy use attributed to the ECM

- Measurement periods range from short-term to continuous
- Additional parameters not selected for field measurement are estimated

Option B. Retrofit Isolation: All Parameter Measurement
- Savings determined by field measurement of the *energy use of the entire system* impacted by the ECM
- Measurement periods range from short-term to continuous depending on the expected variations in the savings due to operating conditions, seasonality, etc., and the length of the reporting period

Option C. Whole Facility
- Savings are determined by measuring energy use for an entire facility or defined facility sub-area.
- Continuous measurements of the entire facility or defined facility sub-area energy use are collected for the reporting period

Option D. Calibrated Simulation
- Detailed energy modeling is used to determine energy savings of the entire facility or defined facility sub-area.
- Simulation routines are demonstrated to adequately model actual energy performance measured in the facility
- Computer simulation modeling expertise is required to employ this option

Energy Service Companies
Energy Service Companies (ESCOs) are commercial business or non-profit firms that offer energy efficiency services across a broad array of competencies, such as:
- Designs and implementation of energy savings projects
- Energy Audits
- Commissioning
- Cost-benefit analysis
- Engineering
- Energy procurement
- Project management
- Risk management

ESCO expertise can vary widely.

ENERGY SAVINGS PERFORMANCE CONTRACTING

An **energy performance contract** is a binding agreement in which an energy services company (ESCO) recommends and implements energy cost measures that are often funded by the resulting project cost savings. Savings may be guaranteed by the ESCO that has assumed risk to realize a return on investment (ROI) based on energy saving calculations.

A typical performance contract consists of the following parts:

- An ESCO conducts all audits, designs all solutions, and provides equipment and all services required including post-installation M&V
- Project financing—long-term project financing provided by a third-party financing company
- Project Savings Guarantee—the ESCO provides a guarantee that the project savings will adequately cover project financing
- A Measurement and Verification (M&V) plan is executed to confirm that ECMs are performing per the guarantee.

SHARED SAVINGS CONTRACTS

A **shared savings contract** is a financing agreement in which an investment firm (or ESCO) authorizes the implementation of ECMs, including retrofits or major capital improvements, in exchange for a percentage of cost savings.

- The financing company provides funding for initial equipment purchases and installation
- Customer pays a percentage of initial costs through energy savings per month or other terms
- Once the financing balance is paid off, the customer gains the full benefits of the energy cost savings

Benefits of Shared Savings Contracts
- Customers do not require up-front capital
- Customer risk is largely mitigated
- Customer and financing firm are both incented to maximize project savings

UTILITY FINANCING

Utility Financing of energy efficiency projects is offered in various regions through a number of methods, including:

- Direct payments on consumer's utility bills to repay financing for energy-related improvements
- Sale of renewable energy credits (RECs), such as solar RECs (SRECs) on the open market to help finance renewable energy projects

Utility Financing Benefits:
- Helps make energy efficiency affordable to more rate payers by expanding access to capital
- Helps utilities to meet their energy demand and usage goals
- Removes the burden for rate payer end-user customers to use existing financial resources
- May address barriers to ECM implementations such as renter/owner split of utility incentives or tax beneficiaries

DEMAND-SIDE MANAGEMENT

Energy efficiency project financing as well as state and local energy reduction incentives may be offered to meet demand-side management objectives. Financing programs to address demand-side management of local areas may be justified due to:

- Patterns of high grid utilization during peak demand hours
- Impact of power plant closures
- Real estate development of commercial and residential buildings

CONTRACTING AND LEASING

Leasing is a financing method generally geared for smaller projects. Advantages:

- Payments align with cost savings: Lease payments can match increase in cash flow from ECM implementation
- Fixed payments: based on estimated savings; payments not based

on performance nor verified with M&V practice

- Tax, depreciation benefits: capital lease or operating lease flexibility to match terms and benefits that best serve end-customer needs
- Shorter leasing terms: shorter terms (5-10 years, e.g.) reduce risk from energy price fluctuations

RISK ASSESSMENT

The **risk assessment** of an energy efficiency project is a review of factors that determine a probability to successfully achieve projected cost savings and corresponding ROI. The assessment can form the basis for project financing and implementation approval.

In addition to capital, installation, and on-going maintenance costs, factors examined to assess risk include:

- Short- and long-term energy fuel pricing projections
- Project design
- Energy baseline assumptions
- Project savings projections
- Energy modeling of the ECM
- Implementation issues
- Using emerging technologies with limited deployments
- Quantifying areas of uncertainty

LOANS, STOCKS AND BONDS

Stock and bond funds have been established to finance energy efficiency projects, via state agencies or directly to the marketplace such as:

- State clean energy funds (CEFs)
- State green banks: provide low-cost, long-term loans for clean energy projects

Bond financing is similar to municipal infrastructure project financing via issue of tax-exempt and taxable bonds. Examples:

- Revenue bonds from state energy for loan financing of energy efficiency improvement projects
- Bond issues for renewable energy project development against individual project cash flows such as a power purchase agreement

SAMPLE QUESTIONS

1. With a "true lease," a tax paying facility will never own the equipment, and thus, they can deduct the full annual lease cost as an operating expense for the year.
    a. True
    b. False

2. Which of the following methods are considered to be common alternative financing methods?
    a. Bonds
    b. Leases
    c. Lotteries
    d. a and b
    e. All of the above

3. A company that does not have a strong maintenance staff and management support for maintenance should consider which of the following financing methods?
    a. Direct Purchase (in-house or borrowed money)
    b. Performance Contracting
    c. Capital Lease
    d. a and c
    e. None of the above

4. All other things equal, if the loan rate is less than a company's minimum attractive rate of return (MARR), the company should:
    a. Not borrow any money and use a true lease
    b. Use a capital lease to delay the purchase
    c. Look only at performance contracting
    d. Borrow a significant percentage of the equipment price if the lender and the balance sheet allow it.

5. Depreciation:
    a. Is an important cash flow
    b. Is used to calculate taxes but is not a cash flow
    c. Is often negative as assets grow in value
    d. Is not important in pre-tax cash flow analysis
    e. b and d

SOLUTIONS TO SAMPLE PROBLEMS

1. a
2. d
3. b
4. d
5. e

Appendix

ENERGY UNIT CONVERSIONS

Electricity

Power
1 kW = 1,000 Watts
1 horsepower (hp) = .746 kW
1 Boiler-horsepower (BHP)
          = 33,745 Btu/hr
          = 9.89 kW
1 watt (W) = 1 Joule/sec (J/sec)
Energy
1 kWh = 3,412 Btu
1 kWh = .03412 therms
1 kWh = .003413 MCF

Pressure

psi = lb/in^2
psig = lb/in^2 gauge measurement
psia = lb/in^2 gauge measurement plus
atmospheric pressure

1 psia = 14.7psi + psig
1 Bar = 14.50 psig
1 psi = 144 lb force/ft^2

Air Volume

1 ft^3 = 0.07788 lbs at 50degF
1 ft^3 = 0.07640 lbs at 60degF
1 ft^3 = 0.07495 lbs at 70degF

Lighting

1 foot-candle = 1 lumen/ft^2
              = approx. 10 lux

Heat Content

1 Btu heats 1 lb water (liquid) by 1°F

1 ton = 12,000 Btu/hr
1 Btu (59 °F) = 1054.80 J
1 kWh = 3.6 x 10^6 J
1 quad = 1x10^{15} Btu

Natural Gas

Energy
1 therm = 100,000 Btu
       = 29.31 kWh
1 dekatherm (Dtherm)= 10 therms
             = 1,000,000 Btu
             = 1 MMBtu
             = 1,000 MBtu
Volume
Volume $_{therm}$ = 100 ft^3
          = 1 CCF
Volume $_{Dtherm}$ = 1000 ft^3
          = 10 CCF

Water, Steam

1 Gallon water = 8.345 lbs
970 Btu heats 1 lb water (liquid) at 212°F to
change state to steam
970 Btu is released from 1 lb steam (vapor)
at 212°F to change state to water (liquid)
144 Btu heats 1 lb ice (solid) at 32°F to
change state to water (liquid)
144 Btu is released from 1 lb water (liquid)
at 32°F to change state to ice (solid)

Water Flow:
500 LB/hr water = 1 gpm water

Degree Days
Heating Degree Days (HDD) = (65°F – avg. Temp. of period) x (days of the given period)
Cooling Degree Days (CDD) = (avg. Temp of period – 65°F) x (days of the given period)

Approximate Heating Values of Common Fuels
Natural Gas   100,000 Btu/therm = 23,600 Btu/lb
Propane LPG   92,500 Btu/gal
Propane gas 60°F   2,500 Btu/ft^3 = 21,000 Btu/lb

#1 Fuel Oil   137,400 Btu/gal
#2 Fuel Oil   139,600 Btu/gal
#3 Fuel Oil   141,800 Btu/gal
#4 Fuel Oil   145,100 Btu/gal
#5 Fuel Oil   148,800 Btu/gal
#6 Fuel Oil   152,400 Btu/gal
Diesel Fuel   139,000 Btu/gal

Methane   1,000 Btu/ft^3
Landfill gas   500 Btu/ft^3
Butane   3,200 Btu/ft^3 = 130,000 Btu/gal (liquid)
Methanol   57,000 Btu/gal
Ethanol   84,400 Btu/gal
Crude Oil   5,100,000 Btu/barrel
Kerosene   135,000 Btu/gal
Waste oil   125,000 Btu/gal
Biodiesel   120,000 Btu/gal (waste vegetable oil)
Gasoline   125,000 Btu/gal = 20,000 Btu/lb

Coal   12,500 Btu/lb = 25,000,000 Btu/ton
Hard Coal (anthracite)   13,000 Btu/lb = 26,000,000 Btu/ton
Soft Coal (bituminous)   12,000 Btu/lb = 24,000,000 Btu/ton

Rubber – pelletized   16,000 Btu/lb = 32,000,000 to 34,000,000 Btu/ton
Plastic   18,000 to 20,000 Btu/lb
Hydrogen   61,000 Btu/lb
Nuclear fission   33,000,000,000 Btu/lb

Softwood   2,000 to 3,000 lb/cord = 10,000,000 to 15,000,000 Btu/cord
Hardwood   4,000 to 5,000 lb/cord = 18,000.000 to 24,000,000 Btu/cord

Sawdust – green   10-13 lb/ft^3 8,000 to 10,000,000 Btu/ton
Sawdust – kiln dry   8 to 10 lb/ft^3 = 14,000,000 to 18,000,000 Btu/ton
Chips (45% moisture)   10 to 30 lb/ft^3 = 7,600,000 Btu/ton
Hogged   10 to 30 lb/ft^3 = 16,000,000 to 20,000,000 Btu/ton
Bark   10 to 20 lb/ft^3 = 9,000,000 to 10,500,000 Btu/ton
Wood pellets (10% moisture)   40-50 lb/ft^3 = 16,000,000 Btu/ton

Corn (shelled)   7,800 to 8,500 Btu/lb
Corn cobs   8,000 to 8,300 Btu/lb

Geometry Formulas

Volume of Cylinder (tank) = $\pi r^2 h$
Area of Cylinder (tank) = $2\pi r^2 + (2\pi r \times h)$
Volume of Sphere = $4/3\pi\ r^3$
Area of Sphere = $4\pi\ r^2$

CHAPTER 1:
CODES AND STANDARDS AND INDOOR AIR QUALITY

Air Change Rates/Hour (ACH)

See pages 242-243.

CHAPTER 2:
ENERGY ACCOUNTING AND ECONOMICS

Discounted After-Tax Cash Flow (ATCF)

ATCF = Annual profit – [(Annual profit – Annual depreciation) x tax rate)]
Straight-line Depreciation = Total Cost of Project/life of project years

Benefit Cost Ratio

$$\text{Benefit Cost Ratio (BCR)} = \frac{\text{Total Benefits (\$)}}{\text{Total Costs (\$)}}$$

Discounted Benefit Cost Ratio

$$\text{BCR} = \frac{\text{PV (Benefits)}}{\text{PV (Costs)}}$$

Air Change Rates/Hour (ACH) (*Continued*)

All spaces in general	4 (minimum)
Assembly halls	4 - 6
Attic spaces for cooling	12 - 15
Auditoriums	8 - 15
Bakeries	20 - 30
Banks	4 - 10
Barber Shops	6 - 10
Bars	20 - 30
Beauty Shops	6 - 10
Boiler rooms	15 - 20
Bowling Alleys	10 - 15
Cafeterias	12 - 15
Churches	8 - 15
Classrooms	6 - 20
Club rooms	12
Clubhouses	20 - 30
Cocktail Lounges	20 - 30
Computer Rooms	15 - 20
Court Houses	4 - 10
Dance halls	6 - 9
Dental Centers	8 - 12
Department Stores	6 - 10
Dining Halls	12 -15
Dining rooms (restaurants)	12
Dress Shops	6 - 10
Drug Shops	6 - 10
Engine rooms	4 - 6
Factory buildings, ordinary	2 - 4
Factory buildings, fumes & moisture	10 - 15
Fire Stations	4 - 10
Foundries	15 - 20
Galvanizing plants	20 - 30
Garages repair	20 - 30
Garages storage	4 - 6
Homes, night cooling	10 - 18
Hospital rooms	4 - 6
Jewelry shops	6 - 10
Kitchens	15 - 60
Laundries	10 - 15
Libraries, public	4
Lunch Rooms	12 -15
Luncheonettes	12 -15

Air Change Rates/Hour (ACH) (*Concluded*)

Nightclubs	20 - 30
Machine shops	6 - 12
Malls	6 - 10
Medical Centers	8 - 12
Medical Clinics	8 - 12
Medical Offices	8 - 12
Mills, paper	15 - 20
Mills, textile general buildings	4
Mills, textile dye houses	15 - 20
Municipal Buildings	4 - 10
Museums	12 -15
Offices, public	3
Offices, private	4
Photo dark rooms	10 - 15
Pig houses	6 - 10
Police Stations	4 - 10
Post Offices	4 - 10
Poultry houses	6 - 10
Precision Manufacturing	10 - 50
Pump rooms	5
Residences	1 - 2
Restaurants	8 - 12
Retail	6 - 10
School Classrooms	4 - 12
Shoe Shops	6 - 10
Shopping Centers	6 - 10
Shops, machine	5
Shops, paint	15 - 20
Shops, woodworking	5
Substation, electric	5 - 10
Supermarkets	4 - 10
Swimming pools	20 - 30
Town Halls	4 - 10
Taverns	20 - 30
Theaters	8 - 15
Transformer rooms	10 - 30
Turbine rooms, electric	5 - 10
Warehouses	2
Waiting rooms, public	4
Warehouses	6 - 30

$$NPV = PV - \text{Initial Project Cost}$$

$$ROI = \frac{\text{Net Gain}}{\text{Total Cost}} = \frac{(\text{Total Benefit} - \text{Total Cost})}{\text{Total Cost}}$$

CHAPTER 3:
ENERGY AUDITS AND INSTRUMENTATION

Energy Indexes
Energy Use Intensity (EUI) = Btus/ft^2 per year
Energy Cost Index (ECI) = $/ft^2 per year

CHAPTER 4:
ELECTRICAL SYSTEMS

Ohm's Law
Power (Watts) = Current (Amps) x Resistance (Ohms)

Power, Voltage, and Current Calculations
Y Circuit
$V_{Line} = V_{phase}$ x 1.73   where
V_{Line} = Voltage from line to line, measured between any two line conductors in a balanced threephase system.
V_{phase} = Voltage measured across any one component in a balanced threephase system.

Delta Circuit
$I_{Line} = I_{phase}$ x 1.73   where
I_{Line} = Current through any one line conductor
I_{phase} = Current through any one component

Load Factor
Load Factor = $\dfrac{\text{Average Load}}{\text{Peak Load}}$ *for a specified period* (month, hours, day(s), etc.)

Power Factor

$$\text{Power Factor} = \frac{\text{Real Power}}{\text{Total Power}} = \cos \Phi$$

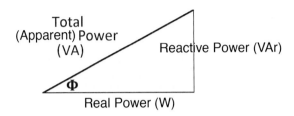

$$(\text{Total Power})^2 = (\text{Real Power})^2 + (\text{Reactive Power})^2$$

Single Phase Power Factor

$$\text{Power Factor} = \frac{\text{Real Power (W)}}{\text{Total Power (V x A)}}$$

$$\text{Real Power (W)} = \text{Total Power (V x A)} \times \text{PF} = \text{V} \times \text{A} \times \text{pf}$$

Three-phase Power Factor

$$\text{Real Power} = \sqrt{3} \times \text{V} \times \text{A} \times \text{PF}$$
$$= 1.73 \times \text{V} \times \text{A} \times \text{PF}$$

Capacitance

$$\text{C (farads)} = \text{KVAR}/2\pi(\text{Frequency in Hz})\text{V}^2$$
$$= \text{KVAR}/377\text{V}^2$$

CHAPTER 5:
HEATING, VENTILATION, AND AIR CONDITIONING (HVAC)

Coefficient of Performance (COP)

$$\text{COP} = \frac{\text{Capacity}}{\text{Load}} = \frac{3.517 \times \text{Tons}}{\text{kW}}$$

Air Conditioning EER or SEER

$$EER = \frac{12 \times \text{Tons (Btu)}}{kW} = COP \times 3.412 \text{ Btu/Wh}$$

$$LOAD \text{ (W)} = \frac{\text{Btu/hr}}{\text{Btu/Wh}} = \frac{\text{Btu/hr}}{EER}$$

$$Energy \text{ (Wh)} = \frac{\text{Load (W)} \times \text{Hrs of Operation}}{COP}$$

Water Flow Rate (gallons per minute) in Heating and Chiller Systems

$q \text{ (Btu/hr)} = 500 \times GPM \times \Delta T \text{ (°F)}$

Flow Rate through a Valve

$GPM = Cv \sqrt{(\Delta P/G)}$ where

    Cv = Valve Sizing Coefficient (unique for each style and size of valve)

    ΔP = Pressure differential (psi)

    G = Specific Gravity of Fluid (water at 60F = 1.000)

Steam Enthalpy with Mass Flow Rate

$q \text{ (Btu/h)} = M_{flow} \text{ (lb/hr)} \times \Delta h$ where

    Δh is the change in enthalpy (h2-h1)

Specific Gravity of Water

See opposite.

Air: Heat transfer sensible heat gain

$q_{sensible} \text{ (Btu/hr)} = CFM(ft3/min) \times \Delta T \text{ (°F)} \times 1.08$ where

    $\Delta T = (T_{\text{outside air dry bulb temp}} - T_{\text{dry bulb temperature air removed}})$

Air: Heat transfer change in Enthalpy

$q_{total} \text{ (Btu/hr)} = CFM \times 4.5 \times \Delta h$ where

    $\Delta h = (h_{\text{outside air enthalpy}} - h_{\text{enthalpy of air removed}})$

Specific Gravity of Water

Water - Specific Gravity

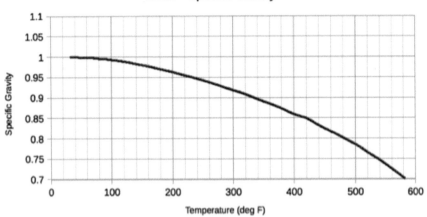

Waste Heat Recovery

q (Btu/h) = M_{flow} (lb/hr) x Cp (Btu/lb-°F) x Δ T (°F) where
Cp is specific heat of the medium

Ventilation Rates – Building Zone Outdoor Air Flow

$V_{bz} = R_p P_z + R_a A_z$ zone—where

V_{bz} = breathing zone outdoor airflow
P_z = zone population
R_p = people outdoor airflow rate required per person from table
R_a = area outdoor airflow rate per unit area from table
A_z = zone floor area

CHAPTER 6:
MOTORS AND DRIVES

Motor Load Factors

$$kW = \frac{NPHP \times (.746\ kW/HP) \times (Load\ Factor)}{Efficiency}$$ where

NPHP is nameplate horsepower

Motor Speed

$$S_{synchronous} (RPM) = \frac{120 \times \text{Freq. (Hz)}}{\text{No. Motor Poles}}$$

# of Poles	Synchronous Speed
2	3600
4	1800
6	1200
8	900

Motor Load Calculation: Slip Method

$$\% \text{ Motor Load} = \frac{(S_{synchronous} - S)}{(S_{synchronous} - S_{Full Load})} = \frac{\text{Time Slip}}{\text{Design Slip}} \quad \text{where}$$

S = measured motor speed, RPM

$S_{synchronous}$ = Motor's synchronous speed (no load) = NLRPM

$S_{Full Load}$ = Motor's full load (rated load) speed = FLRPM

$\text{Slip} = (S_{synchronous} - S)$

Voltage Compensated Slip Method

$$\% \text{ Motor Load} = \frac{(S_{synchronous} - S)}{(S_{synchronous} - S_{Full Load}) \times (V_{rated} / V_{measured}^2)} \quad \text{where}$$

$V_{measured}$ = average RMS line-line voltage

Voltage Compensated Current Ratio

$$\% \text{ Motor Load} = (I_{measured} / I_{rated}) \times (V_{measured} / V_{rated})$$

Direct kWh Calculation

$$\% \text{ Motor Load} = \frac{\sqrt{3} \times V \times I \times PF}{\text{Motor Power}_{rated}}$$

(if not measured directly from a meter)

Heat Load—Motors

Single Phase

Btu/hr = Power (kW) x (Use Factor) x PF x 3412 Btu/kWh

Three-Phase

Btu/hr = Voltage x Current per phase x 1.732 x PF (kW) x (Use Factor) x 3412 Btu/kWh

CHAPTER 7:
INDUSTRIAL SYSTEMS

CHAPTER 8:
BUILDING ENVELOPE

Heat Transfer—Insulation

Thermal Resistance

$$R \text{ Value (ft}^2 \bullet hr \bullet {}^\circ F/Btu) = \frac{d}{K} \quad \text{where}$$

K is coefficient of thermal conductivity of the material
d is thickness of material

Conductance:

$$U \text{ (Btu/ft}^2 \bullet hr \bullet {}^\circ F) = \frac{1}{R}$$

Heat Load/Transfer – Conduction

Q_{total} (Btu/hr) = U x A x ΔT (°F)   where
ΔT = temperature difference between adjacent space and room temp

Q_{total} (Btu/hr) = U x Area$_{surface}$ x (DD/hrs in period) x 24 hr/day, where
DD = total of HDDs or CDDs for the given time period

Heat Transfer
Pipes

$$Q_{\text{Pipe length}} \, (\text{Btu}/\text{hr} \bullet \text{ft}) = \frac{2\pi \times K \times \Delta T(°F)}{\log_n (r_{\text{outside}}/r_{\text{inside}})} \quad \text{where}$$

ΔT = temperature inside pipe – temperature outside pipe
$\quad r$ = radius of pipe

Tanks

$R_{\text{total}} = R_{\text{tank}} + R_{\text{surface coefficient}}$

$$R_{\text{tank}} = \frac{d}{K} \qquad \text{where } d = \text{thickness}$$

Heating and Cooling Load: Solar Radiation
H (Btu/Hr)= U (Btu/Hr ft^2 °F) x Area ft^2 x (CLTD)
  H = Sensible heat gain
  U = Thermal Transmittance for roof or wall or glass.
  A = area of roof, wall or glass calculated from building plans (ft^2)
  CLTD = Cooling Load Temperature Difference (in °F) for glass.

Radiant Sensible Loads (transparent/translucent elements)
H (Btu/Hr) = Area ft^2 x (SHGC) x (SC) x (CLF) where
  H = Sensible heat gain (Btu/Hr)
  A = area of roof, wall or glass calculated from building plans (ft^2)
  SHGC = Solar Heat Gain Coefficient
  SC = Shading Coefficient
  CLF = Solar Cooling Load Factor.

Cooling Load
  q (Btu/hr) = Σ Surfaces [Area$_{\text{surface}}$ (ft^2) x SC x MSHG (Btu/hr-ft^2) x CLF] —where
  SC = Shading Coefficient
  MSHG = maximum solar heat gain (Btu/hr ft^2)
  CLF = cooling load factor

Note:
- R may be expressed as:
  $R = hr \cdot ft^2 \cdot °F/Btu \cdot inch$ per inch $= d/K$
  $R = 1/K$ when thickness is per inch (per 1 inch)

- Reference K table to find $1/R = K$ and the material.

CHAPTER 9:
COMBINED HEAT AND POWER AND RENEWABLE ENERGY

CHAPTER 10:
ENERGY PROCUREMENT

Demand Load (utility billing method using minimum required pf)

$$\text{Billed Demand (kW)} = \text{Actual Demand} \times \frac{\text{Base Power Factor}}{\text{Actual Power Factor}}$$

CHAPTER 11:
BUILDING AUTOMATION AND CONTROL SYSTEMS

CHAPTER 12:
GREEN BUILDINGS, LEED, AND ENERGY STAR

CHAPTER 13:
THERMAL ENERGY STORAGE SYSTEMS

Thermal Storage Formula

$C (Btu_{stored}) = M(lbs) \times Cp \times \Delta T (°F)$ where

Cp is the specific heat capacity of the substance (Btu/lb-°F)

CHAPTER 14:
LIGHTING SYSTEMS

$Lux = \text{Total Lumens} \div \text{Area}$
$1 \text{ Lux (lx)} = 1 \text{ Foot-candle (fc)} \times 10.76$

Ballast Factor

$$BF = \frac{\text{Lamp lumen output on commercial ballast}}{\text{Lamp rated light output}}$$

Average Maintained Illumination

$$\text{Foot Candles} = \frac{\text{Fixtures x Lamps/Fixture x Lumens/Lamp x CU x LLF}}{\text{Room Area (ft}^2)}$$

where
LLF = light loss factor
CU = coefficient of utilization

Lamp Burnout Factor = 1 –% Lamps Allowed to Fail without replacement

Zonal Cavity Design Method

$$\text{Number of Lamps} = \frac{\text{Foot-candles x Room Area (ft}^2)}{\text{Lumens/Lamp x } D_{\text{lamp}} \text{ x CU}}$$

where
D_{lamp} = depreciation factor for the lamp and fixture

Inverse Square Law (Luminance at a specific point)

$$\text{Luminance (lumens/area)} = \frac{\text{Candelas (luminous intensity)}}{(\text{Distance from light source})^2}$$

Room Cavity Ratio (rectangular rooms)
RCR = 5 x MH x [(L + W)/(L x W)] where
MH = mounting height = distance from bottom of fixture and
to workplane

Room Cavity Ratio (irregular shaped rooms)

$$RCR = \frac{2.5 \text{ x (ceiling height) x (perimeter distance)}}{\text{Room Area}}$$

Calculating Number of Lamps, Fixtures, Spacing

$$\text{Spacing Between Fixtures} = \sqrt{\frac{\text{Room Area}}{\text{Required \# Fixtures}}}$$

$$\text{Fixtures per Row (Nrow)} = \frac{\text{Room Length}}{\text{Spacing}}$$

$$\text{Fixtures in Continuous Row} = \frac{\text{Room Length}}{\text{Fixture Length - 1}}$$

$$\text{Fixtures per Column (Ncolumn)} = \frac{\text{Room Width}}{\text{Spacing}}$$

$$\text{Spacing Rows} = \frac{\text{Room Width}}{\text{Fixtures/Row - 1/3}}$$

$$\text{Spacing Columns} = \frac{\text{Room Width}}{\text{Fixtures/Column - 1/3}}$$

Impact on designed light level:

$$\% \text{ Design Light Level} = \frac{\text{Actual \# Fixtures}}{\text{Calculated \# Fixtures}}$$

Heat Load—Lighting

Q (Btu/Hr) = K(kW) x 3412 Btu/kWh where
K is the lighting load in kW

Q (Btu/Hr) = 3.412 Btu/Watt x Watts $_{\text{lamps}}$ x F_{UT} x F_{BF} x CLF where
F_{UT} = Lighting use factor, as appropriate
F_{BF} = Ballast factor allowance, as appropriate
CLF = Cooling Load Factor

IESNA Recommended Foot-Candle Light Levels

Room Type	(Foot-Candles)
Bedroom—Dormitory	20-30
Cafeteria—Eating	20-30
Classroom—General	30-50
Conference Room	30-50
Corridor	5-10
Exhibit Space	30-50
Gymnasium—Exercise/Workout	20-30
Gymnasium—Sports/Games	30-50
Kitchen/Food Prep	30-75
Laboratory (Classroom)	50-75
Laboratory (Professional)	75-120
Library—Stacks	20-50
Library—Reading/Studying	30-50
Loading Dock	10-30
Lobby—Office/General	20-30
Locker Room	10-30
Lounge/Break room	10-30
Mechanical/Electrical Room	20-50
Office—Open	30-50
Office—Private/Closed	30-50
Parking—Interior	5-10
Restroom/Toilet	10-30
Retail Sales	20-50
Stairway	5-10
Storage Room—General	5-20
Workshop	30-75

Light Power Densities

	W/ft²		W/ft²
Automotive Facility	0.982	Manufacturing Facility	1.11
Convention Center	1.08	Motel	0.88
Court House	1.05	Movie Theater	0.83
Dining: Bar Lounge/Leisure	0.99	Multi-Family	0.60
Dining: Cafeteria/Fast Food	0.90	Museum	1.06
Dining: Family	0.89	Office	0.90
Dormitory	0.61	Parking Garage	0.25
Exercise Center	0.88	Penitentiary	0.97
Gymnasium	1.00	Performing Arts Theatre	1.39
Healthcare Clinic	0.87	Police/Fire Station	0.96
Hospital	1.21	Post Office	0.87
Hotel	1.00	Religious Building	1.05
Library	1.18	Retail	1.40
Manufacturing Facility	1.11	School/University	0.99
Motel	0.88	Sports Arena	0.78
Motion Picture Theater	0.83	Town Hall	0.92
Multi-Family	0.60	Transportation	0.77
Museum	1.06	Warehouse	0.66
Office	0.90	Workshop	1.20

Lamp Characteristics

Lamp Characteristics

Type	Common ratings (Watts)	Color rendering	Color Temp. (K)	Life (hours)
Compact fluorescent lamps	5–55	good	2,700–5,000	5,000–10,000
High-pressure mercury lamps	80–750	fair	3,300–3,800	20,000
High-pressure sodium lamps	50–1,000	poor to good	2,000–2,500	6,000–24,000
Incandescent lamps	5–500	good	2,700	1,000–3,000
Induction lamps	23–85	good	3,000–4,000	10,000–60,000
LED lamps	17-36	75-80	3,500- 5,000	50,000 +
Low-pressure sodium lamps	26–180	monochromatic yellow color	1,800	16,000
Low-voltage tungsten halogen lamps	12–100	good	3,000	2,000–5,000
Metal halide lamps	35–2,000	good to excellent	3,000–5,000	6,000–20,000
Tubular fluorescent lamps	4–100	fair to good	2,700–6,500	10,000–15,000
Tungsten halogen lamps	100–2,000	good	3,000	2,000–4,000

CHAPTER 15:
BOILER AND STEAM SYSTEMS

Boiler Efficiency

$$\text{Boiler System Efficiency \%} = \frac{\text{Useful Heat Output (Btus)}}{\text{Heat Input from Fuel (Btus)}} \times 100$$

Boiler Blowdown Rate

$$q_{BD} \text{ (kg/hr)} = \frac{q_S \times f_c}{(b_c - f_c)} \quad \text{where}$$

q_S = steam consumption (kg/hr)
b_c = maximum allowable Total Dissolved Solids(TDS) in the boiler water (ppm)
f_c = TDS in feedwater (ppm)

% Flash Steam Generated

$$\text{Flash \%} = \frac{h(f1) - h(f2)}{H_{latent}}$$

h (f1) = specific enthalpy of saturated water at inlet
h (f2) = specific enthalpy of saturated water at outlet
H_{latent} = latent heat of saturated steam at outlet

Excess Air Factor

$$\text{Excess Air Factor (EAF)} = \frac{\text{Mass (kg) Air to Combust 1 kg of Fuel}}{\text{Stoichiometric Air (AF)}}$$

Air-to-fuel ratio of various fuels

Fuel	Phase	AF	$CO_{2\,max}$ wet	$CO_{2\,max}$ dry
Very light fuel oil	liquid	14.27	13.56	
Light fuel oil	liquid	14.06	13.72	
Medium heavy fuel oil	liquid	13.79	14.00	
Heavy fuel oil	liquid	13.46	14.14	
Bunker C	liquid	12.63	16.23	
Generic Biomass (maf)	solid	5.88	17.91	
Coal A	solid	6.97	16.09	
LPG (90 P : 10 B)	gas	15.55	11.65	
Carbon	solid	11.44	21.00	

Combustion Efficiency and Temperature
Example for 9% oxygen
See figure on page 258.

CHAPTER 16:
MAINTENANCE AND COMMISSIONING

Lighting

$$\text{Calendar Lamp Life (Years)} = \frac{\text{Rated Lamp Life (Hours)}}{\text{Annual Hours of Operation (Hours/Year)}}$$

Lamp Burnout Factor = 1 – [% lamps allowed to fail without replacement]

$$\text{Group Relamping Cost} = \frac{\text{(lamp cost + labor cost)}}{\text{relamping interval}} \quad \text{where}$$

relamping interval = a % of rated lamp life

Spot Relamping Average Annual Cost ($) = (Operating Hours/Year ÷ Rated Lamp Life) x (Lamp Cost + Labor cost per replaced lamp) x number of lamps

Combustion Efficiency and Temperature

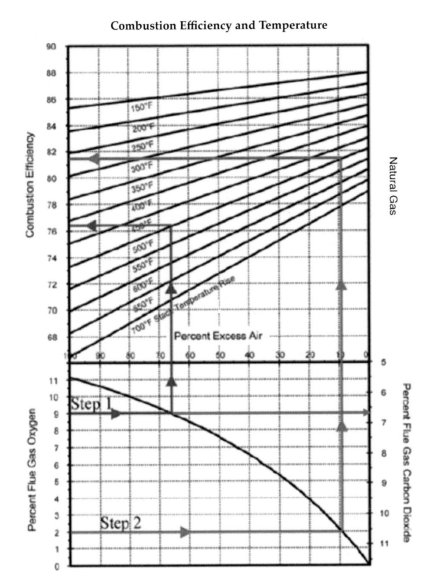

Lighting (*Cont'd*)
Cleaning Cost (\$) = cleaning time / fixture (hrs) x labor rate / hr x number of fixtures

Compressors
Air Compressors

Air Leak Rate (SCFM) = $\dfrac{V \times (P1 - P2)}{T \times 14.7}$ where

   SCFM = standard cubic feet per minute
   V = Volume Air (ft^3)
   P1 = desired system pressure (psig)
   P2 = air pressure (psig) due to leak
   T = time difference (minutes)

Index